REAPING THE BENEFITS OF INDUSTRY 4.0 THROUGH SKILLS DEVELOPMENT IN THE PHILIPPINES

JANUARY 2021

ADB

ASIAN DEVELOPMENT BANK

Contents

Tables, Figures, and Boxes

Boxes

Foreword

Talent and skills are valuable in powering knowledge-based economies. The Fourth Industrial Revolution (4IR) has ushered in extraordinary technological advances, fusing boundaries of physical, digital, and biological worlds to create new paradigms in the way we live, work, and interact. These trends have heralded excitement and fear—excitement in advancing frontiers of human endeavor and fear of negative repercussions on jobs and rising inequalities.

To respond to questions and concerns in developing member countries of the Asian Development Bank (ADB) on how their economies can transition effectively to 4IR, the study *"Reaping the Benefits of Industry 4.0 Through Skills Development in High-Growth Industries in Southeast Asia"* builds an evidence based on opportunities, challenges, and promising approaches in 4IR. It covers Cambodia, Indonesia, the Philippines, and Viet Nam with specific focus on two industries in each country deemed important for growth, employment, and 4IR: tourism and garments in Cambodia, food and beverage manufacturing and automotive manufacturing in Indonesia, information technology and business process outsourcing and electronics in the Philippines, and agro-processing and logistics in Viet Nam.

Much has been written about anticipated loss of millions of jobs arising from automation. At ADB, we take a tempered view. The study reaffirms a positive outlook to 4IR creating new opportunities for quality jobs. While many jobs will indeed be lost as a result of automation, new jobs will emerge through the adoption of technologies that will increase worker productivity and competitiveness of nations, thereby leading to greater prosperity. However, tapping such benefits is predicated on increasing investments in skills development and greater efforts by companies to upskill their workforce to perform new and higher order roles in complementarity with machines.

Adoption of 4IR technologies can increase efficiency and productivity. They enable real-time tracking of supply chains for production and inventory management of raw materials and finished goods. Use of artificial intelligence and machine learning can provide insights into consumer behavior to customize production. Robotic process automation can relieve tedious and repetitive labor-intensive activities, allowing time for higher order functions. Augmented reality and virtual reality can be helpful to train workers in new tasks that they were not familiar with, or skilled in, earlier. Application of 4IR technologies helps developing countries move up the value chain in their products and services. Timely skills development can ensure that automation and artificial intelligence can benefit workers at large.

The study has resulted in a suite of country reports for Cambodia, Indonesia, the Philippines, and Viet Nam, and a synthesis report that captures common elements across the four. They seek to provide policy makers with research and evidence-based solutions for skills and talent development to strengthen the countries' readiness for a transition to 4IR.

The role of governments is crucial in ensuring equitable access to skills development. We expect to see a new balance between physical and virtual workplaces as the gig economy, where employers increasingly rely on part-time freelance workers on short-term contracts, takes firmer position, and widespread digital transformation of citizen services that call for basic digital capabilities in all population groups and rising opportunities for those with advanced digital skills. Job losses will be real, however, a well-prepared 4IR strategy with industry transformation road maps that are recommended in the study can convert disruptions to opportunities to pivot the workforce to new and modern occupations.

The study was completed prior to the coronavirus disease (COVID-19). It is apparent that COVID-19 is accelerating digital transformation. Companies deploying 4IR technologies are likely to recover faster from heavy disruptions arising from the pandemic and be more resilient in the future. Beyond COVID-19, market analysts predict a 'new normal' where digital strategies adopted during the lockdown due to the pandemic will pick up pace. Consumer and producer behavior will most likely be altered permanently with greater digital exposure. The study's recommendations to strengthen widespread digital capabilities, enhance online/distance learning, digital platforms, education technology (EdTech), and simulation-based learning have become more relevant in the aftermath of COVID-19. The study also points to the scope for closer collaboration between public and private sectors, which is also quite relevant in the COVID-19 context. The findings of this study are thus very timely in the discourse to facilitate a sustainable recovery from COVID-19, as countries aspire for accelerating economic diversification and boosting competitiveness using the pandemic as an opportunity for structural reforms.

We welcome your feedback on this report and continued engagement with all stakeholders.

Woochong Um
Director General
Sustainable Development and
Climate Change Department

Ramesh Subramaniam
Director General
South East Asia Department

viii

Preface and Acknowledgments

The ADB study *Reaping the Benefits of Industry 4.0 Through Skills Development in High-Growth Industries in Southeast Asia* marks our effort to bridge research, policy, and practice on the implications of the Fourth Industrial Revolution (4IR) on future job markets. To effectively address this forward-looking topic, the study made use of various sources of secondary information and sought to triangulate information from different primary sources. It included a survey of employers, a survey of training institutions on their readiness for 4IR, and analysis of data from online job portals from each country to assess trends in skills demand. The study used a modeling exercise to estimate job displacement and gains in the selected industries in each of the countries. A review of the policy landscape based on benchmarks from international trends and experiences provides the basis for the action points that countries can use to harness the potential of Industry 4.0 to increase productivity, facilitate skills development, and incentivize industry.

The findings and recommendations from the study point us to collaborate with our partners to implement decisive changes in renewing skills development strategies that acquire a full life cycle approach to skills development. This means that there are no degrees or certificates for life and constant renewals and upskilling are essential. The preponderant focus on institution-based training needs to give way to more flexible and multimodal training to include bootcamps, e-learning, and work-place based training. Training for digital skills at basic, intermediate, and higher levels needs a significant ramp up as workplaces undergo digital transformation.

As co-team leaders, we thank the consultant team led by Fraser Thompson, director, AlphaBeta, for an excellent partnership in this study. The core team in AlphaBeta include Konstantin Matthies, engagement manager; Genevieve Lim, engagement manager; and Richard McClellan, senior advisor. We thank AlphaBeta's national experts Ananto Kusuma Seta (Indonesia), Dao Quang Vinh (Viet Nam), Jose Roland A. Moya (Philippines), and Trevor Sworn (Cambodia). AlphaBeta's team developed the analytical model for the study and collaborated closely with ADB's team to bring new insights and directions and we are grateful for this professional collaboration.

Brajesh Panth, Ayako Inagaki, Robert Guild, and Rana Hasan provided valuable guidance to the study. We thank Shamit Chakravarti, Lynette Perez, Yumiko Yamakawa, and Sakiko Tanaka in ADB's Southeast Asia Human and Social Development Division and Paul Vandenberg and Elisabetta Gentile from the Economic Research and Regional Cooperation Department for providing inputs at various stages of the study and Sophea Mar, Sutarum Wiryono, Vinh Ngo from ADB resident missions in Cambodia, Indonesia, and Viet Nam, respectively, for their valuable support and country-level consultations. Iris Miranda, Sheela Rances, and Dorothy Geronimo from ADB, and Jannis Hoh, Shivin Kohli, and Anna Lim from AlphaBeta provided timely coordination of meetings and activities during the study. We thank April Gallega for coordinating the editing of the reports for publication and Mike Cortes for the cover designs.

The study would not have been possible if not for the leadership of senior government and industry representatives and senior members of the academia in the respective countries. We were heartened to note the high level of interest on the topic of 4IR. In each of the countries, there are already several important initiatives underway to enable industry and companies to move toward application of 4IR. The study was closely coordinated with senior government and industry participants, specifically on the selection of the two sectors for detailed study for each of the countries. The emerging findings of the study were shared in country level workshops. Senior officials and key counterparts consulted are listed at the end of each country report.

We look forward to discussions in taking forward the study's policy recommendations.

Shanti Jagannathan
Principal Education Specialist
Sustainable Development
and Climate Change Department

Sameer Khatiwada
Social Sector Specialist
South East Asia Department

Abbreviations

4IR	Industry 4.0 or Fourth Industrial Revolution
ADB	Asian Development Bank
AI	artificial intelligence
ASEAN	Association of Southeast Asian Nations
BPO	business process outsourcing
EdTech	education technology
F&B	food and beverage
GDP	gross domestic product
IBPAP	IT and Business Process Association of the Philippines
ICT	information and communication technology
ILO	International Labour Organization
IMF	International Monetary Fund
IT	information technology
IT-BPO	information technology and business process outsourcing
ITM	industry transformation map
LFS	labor force survey
MSMEs	micro, small, and medium-sized enterprises
OECD	Organisation for Economic Co-operation and Development
PSA	Philippine Statistics Authority
R&D	research and development
STEP	Skills Measurement Program of the World Bank
TESDA	Technical Education and Skills Development Authority (Philippines)
TVET	technical and vocational education and training
VET	vocational education technology
WEF	World Economic Forum

Executive Summary

Background of the Study

The future of jobs is at the heart of the development conundrum in developing countries in the Asia and Pacific region and preparing the workforce of the future with the right skills and capabilities is central to the technical and vocational education and training (TVET) and skills development portfolio of the Asian Development Bank (ADB). In recent years, the influence of disruptive technologies on jobs and labor markets has intensified worries around extensive job losses arising from automation and the potential disappearance of the comparative advantage of countries based on competitive labor costs. Hence, the readiness of developing countries to effectively address the transition to Industry 4.0 or the Fourth Industrial Revolution (4IR) has become an area of concern. To better understand the implications of 4IR on the future of jobs and to assess the readiness of education and training institutions to prepare for future labor markets, ADB undertook a study that seeks to capture the anticipated transformations of jobs, tasks, and skills and to outline policy directions to prepare the workforce for future jobs.

Scope of the Study

The study covers Cambodia, Indonesia, the Philippines, and Viet Nam and includes the following features:

(i)   It focused on two industries in each country deemed important for growth, employment, and 4IR: tourism and garments in Cambodia, food and beverage (F&B) manufacturing and automotive manufacturing in Indonesia, information technology and business process outsourcing (IT-BPO) and electronics in the Philippines, and agro-processing and logistics in Viet Nam. The table shows the economic importance of each industry in each economy.

(ii) The study includes a survey of employers in the chosen industries, a modeling exercise to estimate job displacement and gains, a survey of training institutions on their readiness for 4IR, and analysis of data from online job portals from each country to assess trends in skills demand.

(iii) The policy landscape was assessed, based on benchmarks derived from international trends and experiences, for its ability to harness the potential of Industry 4.0 to increase productivity, facilitate skills development, and incentivize industry.

(iv) Recommendations suggest how to strengthen policy approaches to 4IR, especially the investments needed for skills and training, new approaches to deliver them, and strategies and actions to enhance the readiness of each country's workforce for 4IR.

The COVID-19 Effect

The study was undertaken and completed prior to the spread of the coronavirus disease (COVID-19), which has caused unprecedented disruptions to labor markets and workforce activities across the world. This study's policy recommendations and strategies to strengthen widespread digital capabilities, enhance online/distance learning, digital platforms, education technology (edtech), and simulation-based learning have become all the more relevant in the aftermath of COVID-19. The key approaches discussed and elaborated in the report are very relevant in the current context of countries experiencing nationwide closures of schools and training institutions. It is also expected that post- COVID-19, there will be operating procedures that constitute a "new normal" that entails far more digital capabilities in the workplace. Hence, the findings of this study and the follow-on policy directions are crucial and very timely for facilitating a sustainable COVID-19 recovery strategy.

The two sectors chosen for the study in the Philippines, IT-BPO and electronics, have been extensively and adversely affected. In the IT-BPO sector, there have been widespread disruptions to business operations due to COVID-19. However, the expectation is that there will be lasting shifts in business practices that embody more digital collaborative tools to support working from home following COVID-19. Similarly, in the electronics industry, recovery after COVID-19 will entail embracing digital supply chains and launching digital sales and marketing initiatives. Hence upskilling and reskilling in 4IR-related occupations is even more urgent for the revival of the economy after COVID-19.

The study obviously does not address the implications of COVID-19 in the Philippines, but the policy directions and future investments for higher-order skills, particularly in the digital domain, are eminently suitable for the country to reimagine new beginnings for the two sectors.

Key Findings

As the study covered four countries in Southeast Asia, a report has been prepared for each of them—Cambodia, Indonesia, the Philippines, and Viet Nam. A synthesis report compiling key findings and a comparative picture across these countries is also available. This report covers the key findings of the study for the Philippines. The IT-BPO and electronics manufacturing industries were selected for an analysis of 4IR in the Philippines in terms of impact on jobs, tasks, and skills. These industries are important for national employment, growth, international competitiveness, and relevance to 4IR technologies. The IT-BPO industry accounted for 2.7% of total employment in 2016, and 6% of gross domestic product in 2015. The Philippines' IT-BPO industry also accounts for 10%–15% of the global IT-BPO market share. The electronics manufacturing industry contributed 2% to total employment, and made up 10.5% of the Philippines' exports in 2018. The study finds that 4IR will have a transformational effect on jobs and skills in these two industries, with great potential for positive gains in jobs and productivity that can be reaped through adequate investments in skills and training. Key findings from the study include:

(i) **4IR will bring both job displacement and job gains.**
   (a) Application of 4IR technologies will lead to a loss of jobs; however, it could also lead to new labor demand, and the study estimates a positive net effect in both IT-BPO and electronics. In both industries, 24% of the current workforce could potentially be displaced by technologies related to 4IR. While the overall patterns of impact in the two industries

are similar, there are some important differences. For example, automation of jobs will be potentially higher for men in the IT-BPO sector, whereas in the electronics manufacturing industry, automation will more likely impact women.

(b) Significant productivity improvements are expected in the two industries with the application of 4IR technologies. The study reports that 60% of IT-BPO employers and 50% of employers in electronics manufacturing expect productivity improvements of more than 25% from the application of 4IR technologies by 2030.

(c) There is already progress underway, with 59% of employers in IT-BPO and 49% of electronics manufacturing employers reporting that they have adopted 4IR technologies in their operations, and the share is expected to increase significantly in both industries by 2025.

(d) The study warns that there are no guarantees that displaced workers can seamlessly move into these new jobs without adequate and timely investments in skills development.

(ii) **Job tasks will shift from routine, physical tasks to higher-order tasks with 4IR.**

(a) The importance of routine physical tasks is expected to decline, with analytical and nonroutine tasks gaining greater attention, which is in line with other studies. In the IT-BPO industry, workers could spend an additional 13.3% of their work week on nonroutine and analytical tasks and 13.3% less on routine physical and interpersonal tasks by 2030. The largest increase in time spent will be for analytical tasks. This is likely driven by how technology will be able to automate responses to simple queries and undertake basic processes, while humans will focus on more complex issues that will test their problem-solving abilities. This could be call center agents responding to a customer query or a unique problem faced by a customer, or information technology (IT) operators dealing with a security breach. On aggregate, a larger task shift is expected for the electronics manufacturing industry compared to the IT-BPO industry. According to the results, workers in the industry could spend an additional 15.8% of their work week on interpersonal and other nonroutine tasks and 15.8% less on routine physical and routine interpersonal tasks.

(b) In terms of skills, evaluation, judgment, and decision-making and numeracy skills will become more important by 2030 in both industries, but the electronics manufacturing industry will also require significant increases in technical skills. There will also be a greater need for advanced technical skills in electronics manufacturing vs. in IT-BPO.

(iii) **Skills shortages and skills levels in both industries need to be addressed.**

(a) While preparing the workforce for 4IR, it is important to address the overall skills shortages and lack of preparation for the workplace. Of those surveyed, 52% of IT-BPO employers and 58% of employers in electronics manufacturing reported that graduates hired in the past year have not been adequately prepared by their prehire education and/or training.

(b) Both industries will demand additional person trainings by 2030, 14.2 million for IT-BPO and 7.5 million for electronics manufacturing. One person training refers to training one worker in one skill from the average level required by corresponding occupation and industry in 2018 to the required level in 2030. In both industries, on-the-job training will be the critical form of skills development. Education and training institutions also need to prepare graduates better for entry-level positions.

(iv) **Training institutions in the Philippines need to prepare for the challenges of 4IR.**

    (a) Encouragingly, there is strong alignment between the skills that training institutions believe will be particularly important because of 4IR and the perceptions of employers in the IT-BPO and electronics manufacturing industries. However, some training institutions may be struggling to keep pace with the rate of change in skills demand. For example, 46% of training institutions surveyed review and update their curricula less than annually, and fewer than half of institutions provide information on job-market conditions to their students.

    (b) Despite close engagement with industry, there is a significant mismatch in the perceptions of skills preparation between employers and training institutions. For example, while 90% of training institutions believe their graduates to be adequately prepared for the job market, only 53% of IT-BPO employers and 58% of electronics manufacturing employers agree.

(v) **Courses and training delivery have begun to change but further transformation is needed.** The study found promising self-reported trends in training institutions of adapting teaching and learning in the classrooms, with greater assimilation of technology, particularly digital training. More than 60% of institutions use online self-learning tools. However, the deployment of advanced technologies is still limited—only 15% have adopted virtual learning platforms. The quality and standards used in these new tools have yet to be ascertained.

(vi) **Despite some promising policy initiatives, greater coordination is needed to adequately prepare the Philippines for 4IR.**

    (a) There are a number of innovative policies related to 4IR currently underway in the Philippines, including the JobStart program; the Small Enterprise Technology and Upgrading Program (SETUP) from the Department of Science and Technology (DOST), which aims to encourage micro, small, and medium enterprises (MSMEs) to adopt technological innovations; and the *JobsFit Labor Market Information Report* from the Department of Labor and Employment (DOLE). However, a lack of coordination across government departments on 4IR and a lack of integration with the skills agenda could hamstring the overall effectiveness of these efforts.

    (b) There is currently no single consolidated 4IR strategy, and national policies very much reside within single government departments. While the Comprehensive National Industry Strategy and Inclusive Innovation Industrial Strategy of the Department of Trade and Industry (DTI) establish the policy for enhancing technology development and research efforts to bring about greater economic productivity, the Technical Education and Skills Development Authority (TESDA) has a separate National Technical Education and Skills Development Plan (NTESDP) that seeks to equip the workforce with the skills to facilitate this transition to 4IR. On the other hand, DOST is leading policies to enhance investments in science and technology education and training. There is much scope for these interrelated policies to be consolidated into a common 4IR road map adopted by the Government of the Philippines.

    (c) There are also specific areas of policy challenges, including a lack of responsiveness of the curriculum to 4IR skills needs, a lack of incentives for employers to invest in the skills development of their workers, and an overly strong emphasis on traditional qualifications attained through the education system or competency assessments—as opposed to past work experiences and the skills gained through them.

Key Recommendations and Way Forward

To address current gaps in policy actions and enhance the effectiveness of implementation mechanisms, seven recommendations have been identified for the Philippines to strengthen its preparedness for 4IR. A multistakeholder approach to the actions in all of these recommendations will be critical for their effectiveness. For each of them, a potential lead (from either the government or the private sector) has been identified, along with a suggested list of stakeholders to be engaged when developing and implementing the recommended actions. These recommended actions include:

(i) **Develop 4IR transformation road maps for key sectors.** While there has been progress in aiming to harmonize approaches, including the recent memorandum of understanding between DTI and DOST for interministerial collaboration on 4IR, it may be practically challenging to develop a cohesive nationwide approach. A three-step approach could be explored to address this: (i) conducting a diagnostic to understand current areas of misalignment; (ii) developing a pilot program for one industry (similar to Singapore's industry transformation maps [ITMs]); and (iii) explore ways of scaling this approach to other industries.

(ii) **Develop a series of industry-led technical and vocational education and training programs targeting skills for 4IR.** To strengthen the quality and relevance of TVET programs, building on existing mechanisms for industry engagement, there could be a focus on developing courses and credentials for 4IR in key industries, including IT-BPO and electronics manufacturing. The McKinsey-founded independent nonprofit Generation is a good example of an industry-led program. Over 30,000 people from 13 countries have graduated from its programs; of these, 81% are employed 3 months after graduation and at salaries two to six times higher than their previous earnings.

(iii) **Explore opportunities to increase curriculum responsiveness.** While there are mechanisms in the country's education system to incorporate curriculum changes to meet industry needs, the speed at which these updates can take place is currently restricted, which is detrimental in a landscape of rapidly evolving technologies and associated skills needs. Working with educational institutions to properly understand the reasons for time lags and find ways to address them in incorporating curriculum changes is recommended.

(iv) **Upgrade training delivery through 4IR technology in classrooms and training facilities.** An effective way of preparing students or future workers for 4IR—or at least equipping them with the basic computer literacy skills necessary to excel in the future economy—is to apply 4IR technologies in classrooms. As shown in the training institution in Chapter 2, however, technology adoption in the classroom in the Philippines is currently limited in many institutions. It is recommended that DOST, TESDA, the Department of Education (DepEd), and even the Commission on Higher Education (CHED) collaborate to explore potential blended learning approaches whereby technology and traditional instruction methods are fused together in a low-cost and scalable way.

(v) **Develop flexible and modular skills certification programs.** It is recommended that the Philippines explores the development of flexible skills certification programs that recognize skills attainment outside traditional education channels. A good example of a skills-based accreditation system is the Malaysian Skills Certification Program, under which skills certificates are granted to workers who do not have any formal educational qualifications but who have obtained relevant knowledge, experience, and skills in the workplace to enhance their career prospects.

(vi) **Implement an incentive scheme for firms to train employees for 4IR.** Despite the substantial productivity gains 4IR technologies could bring about, employer training rates in the country remain low due to a number of market failures relating to information asymmetries around the benefits and availability of training, as well as limited training budgets. It is thus critical to develop a set of support programs to encourage firms to invest in relevant 4IR training for their workers. A starting point would be to develop alternative models and conduct cost–benefit assessments to help understand their feasibility.

(vii) **Formulate new approaches and measures to strengthen inclusion and social protection in the context of 4IR.** Given the currently limited social protection even for regular workers, there is still some way to achieving this for on-demand or flexible workers in the new economy. It is recommended that cost–benefit analyses of several policy options for the social protection of such workers be conducted and potentially pilot schemes developed to test their broader applicability.

While these recommendations apply to both the IT-BPO and electronics manufacturing industries, there are a set of priorities unique to each industry that should be considered when implementing the respective actions. These include:

(i) **Information technology and business process outsourcing industry.** Enhance the effectiveness of current industry training institution engagement efforts; ensure focus on training critical thinking and complex problem-solving skills; and improve the capacity of employers to deliver on-the-job training.

(ii) **Electronics manufacturing industry.** Support 4IR knowledge transfer from large multinational companies to MSMEs; address the potentially disproportionate impact of technological disruption on women; and develop a standardized set of 4IR skills requirements and training quality standards.

CHAPTER 1

The Industry 4.0 Skills Challenge

This chapter investigates the demand and supply of skills driven by Industry 4.0 (4IR) technology adoption for both the information technology and business process outsourcing (IT-BPO) and electronics manufacturing industries in the Philippines. The analysis utilized a range of data, including employer surveys and interviews, online job board data, and national labor market statistics.

In both industries, the impact of 4IR will be transformative for jobs and skills. The analysis shows that despite widespread concerns of significant automation and loss of jobs associated with 4IR, the net impact on jobs for both industries to 2030 is likely to be positive, with more jobs being created than lost. However, there are no guarantees that displaced workers can seamlessly move into these new jobs, as they will likely lack the relevant skills. In both industries, 24% of the current workforce could potentially be displaced by technologies related to 4IR. While the overall patterns of impact in the two industries are similar, there are some important differences. For example, automation of jobs will be potentially higher for men in the IT-BPO industry, whereas in the electronics manufacturing industry, automation will have a greater impact on women.

In terms of skills, evaluation, judgment, and decision-making and numeracy will become more important by 2030 in both industries, but the electronics manufacturing industry will also require significant increases in technical skills. There will also be a greater need for advanced technical skills in electronics manufacturing vs. in IT-BPO.

Both sectors will demand additional person trainings by 2030, 14.2 million from the IT-BPO industry and 7.5 million from electronics manufacuring.[1] In both industries, on-the-job training will be the critical form of skills development.

Industry 4.0 and the Relevance for the Philippines

4IR is a widely used but often misunderstood term that refers to a range of new technologies impacting the workplace. The term was first conceptualized to describe data exchange technologies used in manufacturing. However, it has now acquired a broader meaning (and is sometimes referred to as the "Fourth Industrial Revolution"), where it refers to technologies applied across all sectors that combine the physical, digital and biological worlds.[2] These technologies include (among others) cyber–physical systems, the Internet of Things (IoT), artificial intelligence (AI), cloud computing, and cognitive computing.

---

[1] One person training refers to training one worker in one skill from the average level required by their occupation in their industry in 2018 to the required level in 2030.

[2] K. Schwab. 2017. *The Fourth Industrial Revolution.* https://books.google.com.sg/books?hl=en&lr=&id=ST_FDAAAQBAJ&oi=fnd&pg=PR7&dq=klaus+schwab+fourth+industrial+revolution&ots=DTnvbTqvTQ&sig=aOLqcUCFsLKbNpjWa5kr2Sjzhu4#v=onepage&q=klaus%20schwab%20fourth%20industrial%20revolution&f=false.

4IR is a very different concept from previous industrial revolutions, both in terms of scope and technologies (Figure 1). The first industrial revolution in the 18th century was marked by a transition from hand production methods to machines through the use of steam power and water power. The second industrial revolution occurred in the 19th century and involved the use of extensive railroad networks and the telegraph to allow the faster transfer of people and ideas, combined with factory electrification and the creation of mass production assembly line approaches. The third industrial revolution occurred in the late 20th century and is often referred to as the digital revolution, involving the use of computers, the internet, robots and automation, and electronics manufacturing. 4IR builds on these past industrial revolutions but includes a far broader array of technologies with applicability across all industries. In this regard, it is fundamentally different from the past industrial revolutions in its potential implications for economies and the workforce.

What could 4IR mean for the Philippines? According to an International Labour Organization (ILO) study in 2017, Philippine enterprises perceived the advance of technology as the second biggest economic opportunity in the period to 2025, just after a rise in domestic demand.[3] However, adoption

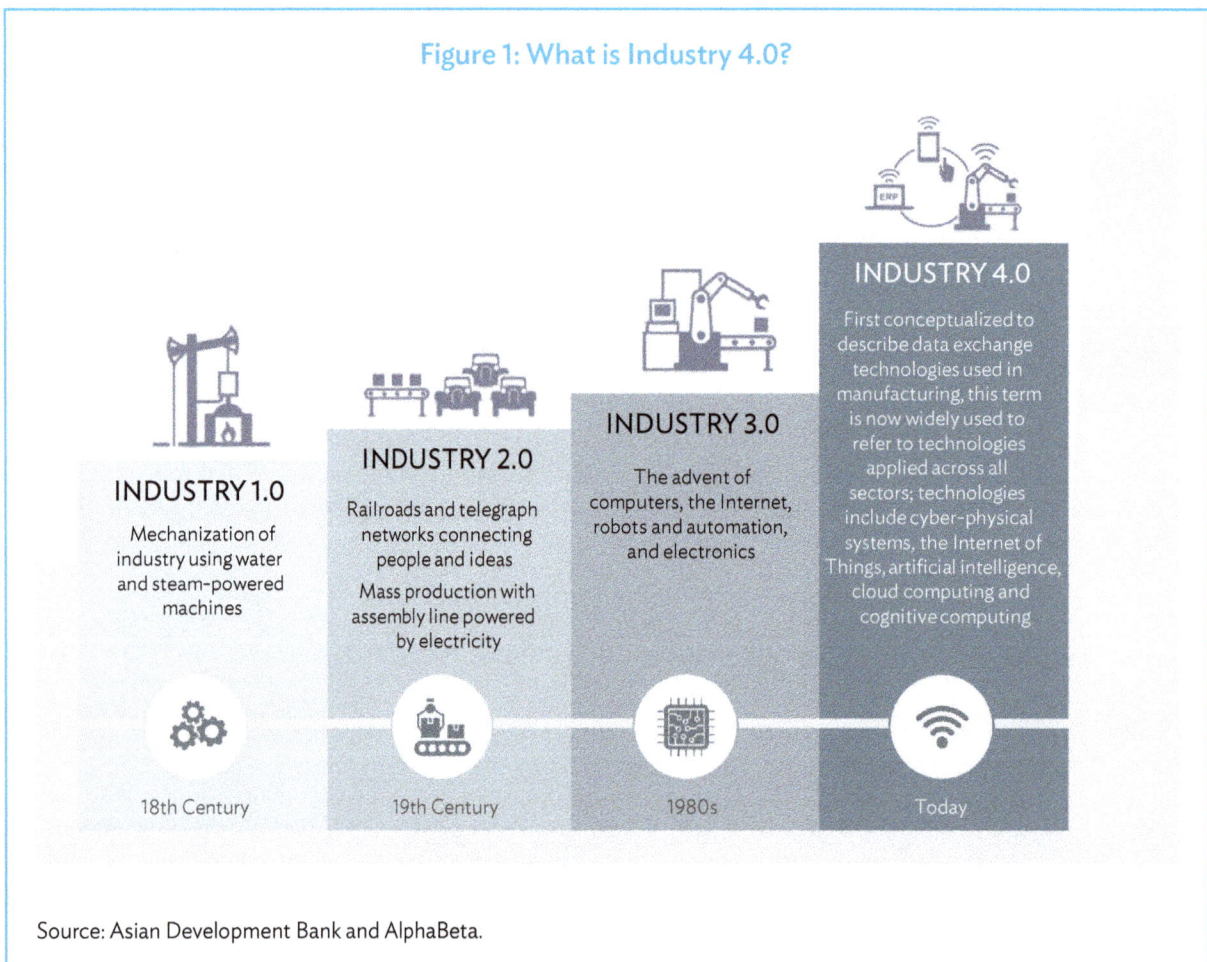

Figure 1: What is Industry 4.0?

INDUSTRY 1.0

Mechanization of industry using water and steam-powered machines

18th Century

INDUSTRY 2.0

Railroads and telegraph networks connecting people and ideas

Mass production with assembly line powered by electricity

19th Century

INDUSTRY 3.0

The advent of computers, the Internet, robots and automation, and electronics

1980s

INDUSTRY 4.0

First conceptualized to describe data exchange technologies used in manufacturing, this term is now widely used to refer to technologies applied across all sectors; technologies include cyber-physical systems, the Internet of Things, artificial intelligence, cloud computing and cognitive computing

Today

Source: Asian Development Bank and AlphaBeta.

---

[3]   Bureau of Employers' Activities & ILO. 2017. *ASEAN in Transformation—How Technology is Changing Jobs and Enterprises —The Philippines Country Brief.* https://www.ilo.org/actemp/publications/WCMS_579667/lang--en/index.htm.

rates of 4IR leave much to be desired. The same study found that only 27% of Philippine enterprises are currently actively upgrading their technology. Costs seem to be a major reason, as over 30% of firms cited high fixed-capital costs as the main barrier to upgrading their technology and another 15% cited high licensing costs. 4IR also holds a large opportunity for the country due to its current underuse of technologies, and hence, large potential room for improvement. For example, the Philippines ranked 54th in the Global Innovation Index 2019. While this is a significant improvement from the previous year's ranking of 73rd, the Philippines is still behind fellow Association of Southeast Asian Nations (ASEAN) member states such as Malaysia (ranked 35th), Viet Nam (42nd), and Thailand (43rd).[4]

With the large potential impact of 4IR technology adoption, there are concerns about the impact on employment. Most concerns revolve around fears that 4IR could lead to mass unemployment as (i) workers are replaced by machines or (ii) workers do not have the right skills to effectively work alongside 4IR technologies or transition into new emerging jobs. According to the ILO, 49% of employment in the Philippines is at high risk of automation (footnote 3). There are also potential gender equity concerns. The probability of occupying a high risk, automatable job is approximately 2.4 times higher for women than for men.

Understanding how the skills landscape is likely to change under 4IR is becoming harder in the face of the rapid pace at which technology is developing and being adopted. This means traditional approaches to assessing skills gaps, often relying on time-intensive processes to collect data that quickly become outdated, may no longer be suitable. This study explores a new approach to understanding the labor market implications of 4IR that tries to address gaps in previous studies. Some of the key design aspects in the methodology include:

(i)   **Use of local data.** This study uses a variety of local data sources, including the Philippines' National Labor Force Survey (LFS), the World Bank's Skills Measurement Program (STEP) survey for the Philippines, as well as surveys of Philippine businesses in the IT-BPO and electronics manufacturing industries.

(ii) **Use of current market information.** Given the rapid change in the labor market, existing labor market surveys can quickly become obsolete. To address this concern, this report uses information on skills profiles for current occupations advertised in online job portals in the Philippines.

(iii) **Focus on supply, not just demand.** Much of the past research has examined only changes in occupations and skills related to 4IR.[5] This study aims to go further by examining the supply landscape, including understanding the volume and types of training required (e.g., on-the-job training, short professional courses, etc.) and conducting a survey of current training institutions in the Philippines to understand the degree to which they currently address the shifts in demand for skills being seen in this analysis.

---

4   Cornell University, INSEAD, and World Intellectual Property Organization. 2019. *Global Innovation Index 2019: Creating Healthy Lives—The Future of Medical Innovation.* https://www.wipo.int/edocs/pubdocs/en/wipo_pub_gii_2019.pdf.

5   A thorough review of the relevant literature can be found in ADB. 2018. *Asian Development Outlook 2018: How Technology Affects Jobs.* https://www.adb.org/publications/asian-development-outlook-2018-how-technology-affects-jobs.

Industry Selection

Two industries were chosen to conduct this analysis of 4IR implications for the demand and supply of skills. A two-step methodology was used to select the industries:

(i) **Shortlisting industries prioritized by the Government of the Philippines for future growth or for 4IR application.** This included reviewing the Philippine Innovative Inclusive Industrial Strategy (I³S), the Philippine Development Plan 2017–2022 and the National Technical Education and Skills Development Plan (NTESDP) 2018–2022.

(ii) **Scoring and ranking shortlisted industries according to a set of criteria:**
(a) How significant is the industry's contribution to the country's employment?
(b) Does it figure strong recent employment growth?
(c) Are its exports internationally competitive?
(d) Is 4IR of relevance to the industry?
(e) Is the relevant data available for the industry analysis?

The industries were then tested with various stakeholders during a country consultation conducted in July 2019. Based on this process, the IT-BPO and electronics manufacturing industries were selected for the analysis:

(i) **Information technology and business process outsourcing.** The IT-BPO industry in the Philippines has grown rapidly since the early 2000s. The industry's annual average growth rates have been estimated at 17%–18%; between 2000 and 2015, BPO's contribution to total gross domestic product increased from less than 1% to 6% (footnote 5). The industry's rapid development has been driven by a large pool of service-minded and English-speaking workforce, supportive government policies, and business associations (footnote 5). According to the IT & Business Process Association of the Philippines (IBPAP), the industry contributed 2.7% to total employment in the Philippines in 2016 and accounted for 10%–15% of global IT-BPO market share. The information and communications technology (ICT) services industry, much of which falls under IT-BPO in the Philippines, is also the most desired industry for students to work in, for both men and women: 25.7% of male and 18.5% of female Philippine students surveyed by the ILO in 2017 expressed a desire to work in this industry, almost double the share of the second most desirable industry (hotels or restaurants). The industry is also facing large potential disruption by 4IR as 89% of Filipino BPO workers are at high risk of automation (footnote 5).

(ii) **Electronics manufacturing.** According to the latest available national statistics, the electronics manufacturing industry contributed 2% to total employment in 2015. According to the United Nations Conference on Trade and Development statistics database, electronics manufacturing made up 10.5% of the Philippines' exports in 2018, being the fifth largest exporter of electronics manufacturing in Southeast Asia and 19th worldwide. The country also has a comparative advantage in both these product groups. The electronics manufacturing industry has been highlighted as a key industry even among the industries prioritized by the government for 4IR technology adoption. From country consultations, it emerged that there is a push by both government and industry to shift industry activities further up the value chain—from parts manufacturing toward fully fledged assembly and production centers. There are also early collaborations between industry, government, and educational institutions in place, but stakeholders voiced a need to understand skilling impacts and priorities.

Information Technology–Business Process Outsourcing Industry

IT-BPO could be transformed by 4IR. Digital technologies have created the market opportunity for IT-BPO and further 4IR technologies (such as chatbots and the use of big data) to greatly improve productivity. However, the labor market disruption could be significant given the advancements in AI-related technologies that can increasingly automate some of these services. This research estimates that 4IR technologies could displace 24% of today's employment in the industry, which represents more than 286,000 workers. This is based on insights from the current tasks performed by workers and how employers expect those tasks to change given the adoption of 4IR technologies (based on the employer survey). Clearly, there is a great deal of uncertainty in some of these future projections.

However, contrary to some perceptions that 4IR will lead to mass unemployment, the research provides an optimistic assessment overall. Net employment from 4IR may actually rise in the IT-BPO industry as displacement effects from 4IR are offset by employment linked to productivity gains, i.e., the income effect. The skills required by workers in this industry will need to change markedly, however, with digital and technical skills, and skills related to critical thinking and adaptive learning, expected to increase significantly in demand. Much of this skills development will need to come from on-the-job training, and in sum, there will need to be 14.2 million additional person trainings[6] by 2030.

Relevance of Industry 4.0

4IR represents both an opportunity and a threat to the IT-BPO industry in many countries. Digital technologies have spurred the growth of IT-BPO and could further increase the productivity of workers by providing real-time insights into customers. However, the increasing sophistication of AI technologies could potentially remove, to a large degree, the need for human involvement in many IT-BPO functions.

There are various 4IR technologies of relevance for the IT-BPO industry. The two key technologies impacting the BPO industry are identified as cloud technology and robotic process automation (footnote 5). Robotic process automation refers to AI algorithms capable of performing both highly structured and ordinary tasks. Some key applications of these technologies include:

(i)   **Chatbots.** A chatbot is a bot that interacts with business customers just like a living person. It can be used as an artificial agent that talks with the customers and can give adequate advice when a live agent is unavailable. This saves the business time and earns customer satisfaction by providing service 24 hours a day, 7 days a week. Past research has shown that healthcare companies and banks that use chatbots to deal with customer queries can save around 4 minutes, or more than 50 cents per inquiry.[7] Recent advances in AI have seen chatbots become difficult to distinguish from humans. Gartner, the research firm, conducted a recent survey of chief information officers; 21% of respondents stated that they plan to deploy some kind of conversational interface in the medium to long term (compared to just 4% today).[8]

(ii) **Big data.** Big data refers to the ability to analyze extremely large volumes of data, extract insights, and act on them closer to real time. Much of this has been facilitated by cloud computing, which allows larger storage and faster computing power without the need for local physical hardware. This has a range of benefits in the BPO industry. Predictive analytics can help

---

6   One person training refers to training one worker in one skill from the average level required by their occupation in their industry in 2018 to the required level in 2030.

7   P. Olson. 2018. Google, Microsoft And Startups Are Going To War On Chatbot Technology, *Forbes.* 27 July. https://www.forbes.com/sites/parmyolson/2018/07/27/google-microsoft-and-startups-are-going-to-war-on-chatbot-technology/#d9229c561b67.

8   Gartner Research. 2018. *Market Guide for Conversational Platforms.* https://www.gartner.com/en/documents/3879492.

Figure 2: Sentiments toward 4IR in the IT-BPO Industry in the Philippines

IT-BPO: 4IR readiness

Companies in the IT-BPO industry appear to have a good understanding of Industry 4.0 but adoption and adoption plans are not advanced

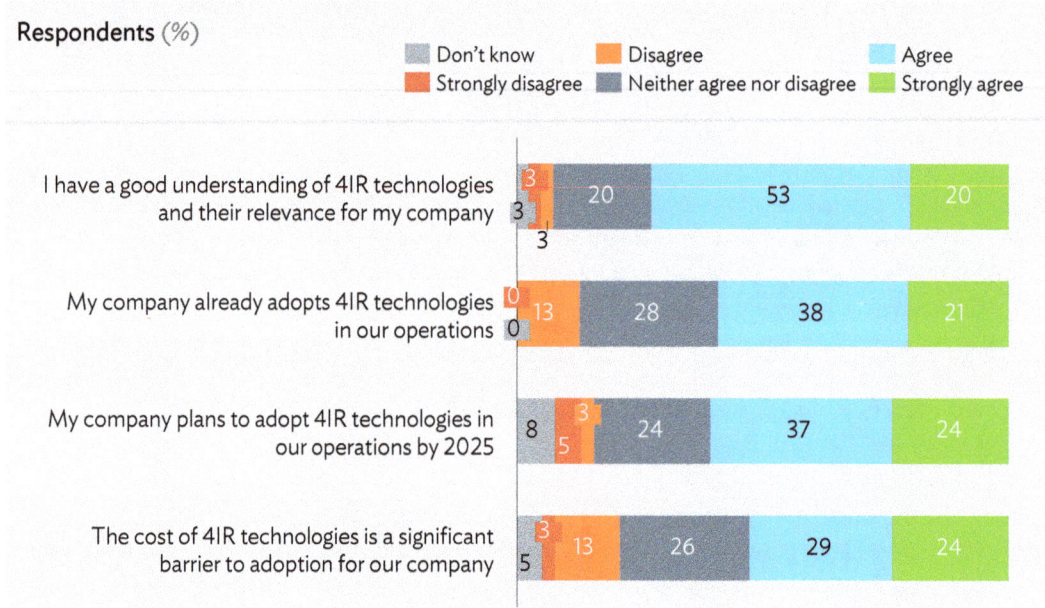

Respondents (%)

Legend: Don't know, Disagree, Agree, Strongly disagree, Neither agree nor disagree, Strongly agree

Statement	Strongly disagree	Disagree	Don't know	Neither agree nor disagree	Agree	Strongly agree
I have a good understanding of 4IR technologies and their relevance for my company	3	1	3	20	53	20
My company already adopts 4IR technologies in our operations	0	13	0	28	38	21
My company plans to adopt 4IR technologies in our operations by 2025	5	3	8	24	37	24
The cost of 4IR technologies is a significant barrier to adoption for our company	5	13	3	26	29	24

4IR = Industry 4.0 or Fourth Industrial Revolution, IT-BPO = information technology and business process outsourcing.
Source: Asian Development Bank and AlphaBeta.

in analyzing customer preferences and increase customer satisfaction. With the information derived from analytics, BPOs can also design targeted programs for customer engagement. Research by Accenture suggests that BPOs can reap significant benefits from adopting these more advanced analytical approaches.[9]

Adoption and understanding of 4IR technologies across the IT-BPO industry in the Philippines is somewhat mixed (Figure 2): 59% of companies surveyed agreed or strongly agreed when asked whether they adopted 4IR technologies in their operations and 73% claimed to have a good understanding of the relevance of 4IR technologies to their industry. Cost remains a critical barrier to adoption: 53% of employers stated that the cost of technology is a significant barrier to adopting 4IR; this aligns with research by the ILO.

More than 60% of employers in the IT-BPO industry in the Philippines anticipated productivity improvements of more than 25% from adopting 4IR over the next 5 years (Figure 3).

9   Accenture. 2012. *Utilizing Analytics to Maximize Business Outcomes.* https://www.accenture.com/us-en/~/media/accenture/conversion-assets/dotcom/documents/global/pdf/technology_7/accenture-utilizing-analytics-maximize-business-outcomes.pdf.

Figure 3: Expected Productivity Improvement Due to 4IR Technologies in 5 Years

IT-BPO: Productivity

Over 60% of employers in the IT-BPO industry expect a productivity increase by more than 25% from 4IR technologies over the next 5 years

Respondents (%)

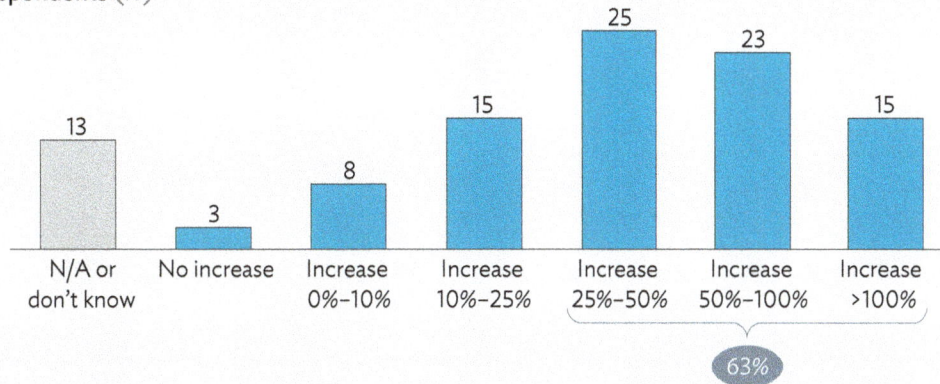

N/A or don't know	No increase	Increase 0%–10%	Increase 10%–25%	Increase 25%–50%	Increase 50%–100%	Increase >100%
13	3	8	15	25	23	15

63%

4IR = Industry 4.0 or Fourth Industrial Revolution, IT-BPO = information technology and business process outsourcing, N/A = not applicable.

Source: Asian Development Bank and AlphaBeta.

Skills Demand Analysis

Employment Implications

The analysis examines two factors influencing employment in the IT-BPO industry related to 4IR:

(i) **Displacement effect.** This refers to the number of jobs lost due to the automation of tasks through the application 4IR technology. Jobs are only lost if the tasks automated by technology make up such a significant proportion of the workers' time spent at work or are so essential to their role that they are no longer needed. The analysis estimates this displacement to be 24% of today's employment, equivalent to more than 286,000 workers.

(ii) **Productivity effect**. Sometimes called a scale effect, this refers to automation improving productivity and lowering production costs. Under normal conditions, this lowers the price of goods and services, which raises demand for them. To the extent that increased demand requires hiring more workers, it could offset the job losses from automation.[10]

Contrary to some perceptions that 4IR will lead to mass unemployment, the research provides an optimistic assessment overall. Net employment from 4IR may actually rise in the IT-BPO industry as job losses from 4IR are offset by employment linked to productivity gains (Figure 4).

---

[10] Automation can also spawn new labor-intensive tasks and jobs, raising demand for labor. New job categories could emerge as 4IR technologies are introduced into production, for example, or when a more sophisticated industrial robot is introduced on a factory floor and needs programming. This is referred to in the literature as the "reinstatement effect." This effect was not estimated in this analysis due to a lack of robust data to size it. Further details on the reinstatement effect are in ADB. 2018. *Asian Development Outlook 2018: How Technology Affects Jobs.* https://www.adb.org/publications/asian-development-outlook-2018-how-technology-affects-jobs.

Figure 4: Modelled Impact of 4IR on Number of Jobs between 2018 and 2030 in the Philippines' IT-BPO Industry

IT-BPO: Jobs

The overall impact of 4IR on jobs is likely to be limited as negative displacement effects are potentially offset by positive income effects

Displacement and income effects of 4IR on jobs, 2018–2030 (%)

Effect	Description	Impact
Displacement	Job reductions due to labor-substitution effects of 4IR	(24)
Productivity	Additional labor demand stimulated by 4IR-enabled productivity gains	35
Net	Combination of displacement and income effects	11

( ) = negative, 4IR = Industry 4.0 or Fourth Industrial Revolution, GDP = gross domestic product, IMF = International Monetary Fund, IT-BPO = information technology and business process outsourcing, LFS = labor force survey, PSA = Philippine Statistics Authority, STEP = Skills Measurement Program of the World Bank.

Note: Change in jobs based on accelerated adoption scenario of 4IR technologies.

Sources: Industry employment -PSA, LFS 2017 and ILO; GDP/Output –PSA and IMF Article IV; STEP survey data; Employer survey on impact of 4IR on the IT-BPO industry in the Philippines, n= 32+; Job portal data: jobs in the IT-BPO industry scraped from the job portal Bestjobs.ph from July to August 2019.

However, even though the overall impact on employment appears to be positive, this does not mean that 4IR could not lead to substantial numbers of workers losing their jobs. There are four critical challenges to realizing the theoretical positive income effect:

(i)   There is no guarantee that the 24% of workers who lose their jobs will be able to seamlessly move into the 35% of jobs created. The transition may not occur if workers cannot be appropriately reskilled.

(ii)   The new jobs may not materialize if there is a lack of suitable skills in the local workforce to support them. In short, the Philippines' approach to skills development will be critical in realizing a positive labor market outcome related to 4IR in this industry.

(iii) There could be potential time lags to the implementation of 4IR, the job losses and the manifestation of productivity benefits. Hence, the productivity gains generating additional income that make new employment possible may take several years to materialize, reducing the positive impact by 2030.

(iv)   Some of the productivity benefits may be absorbed by companies as higher profits if industries are not competitive, or distributed to remaining workers in the form of higher wages if the supply of labor is inelastic. This means that rather than creating additional employment, productivity benefits could generate higher returns for existing actors in the market.

Figure 5: Employers' Expected Impact of 4IR on Relative Number of Jobs by Occupation in the Philippines' IT-BPO Industry, 2019–2025

IT-BPO: Jobs

Employers expect technical customer facing jobs to increase while manual and administrative jobs are expected to decrease due to 4IR

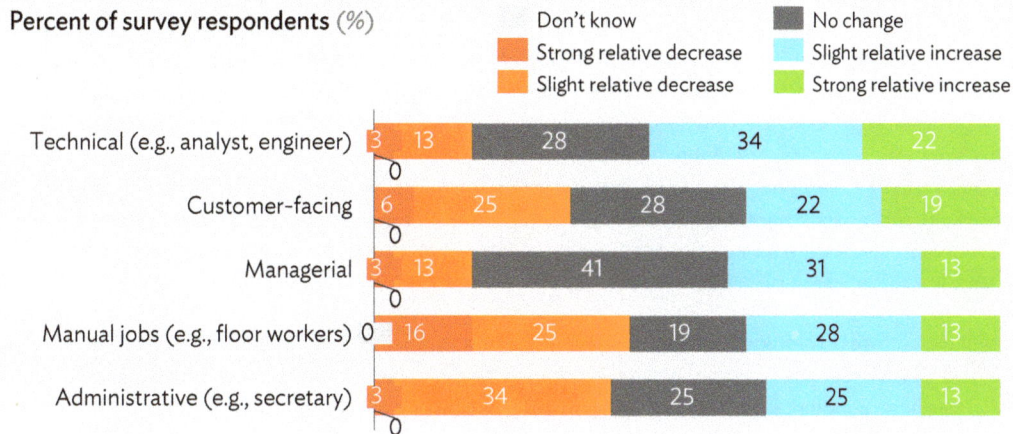

Percent of survey respondents (%)

Legend: Don't know | No change | Strong relative decrease | Slight relative increase | Slight relative decrease | Strong relative increase

Occupation	Strong rel. decrease	Slight rel. decrease	No change	Slight rel. increase	Strong rel. increase	Don't know
Technical (e.g., analyst, engineer)	3	13	28	34	22	0
Customer-facing	6	25	28	22	19	0
Managerial	3	13	41	31	13	0
Manual jobs (e.g., floor workers)	0	16	25	19	28	13
Administrative (e.g., secretary)	3	34	25	25	13	0

4IR = Industry 4.0 or Fourth Industrial Revolution, IT-BPO = information technology and business process outsourcing.

Note: Managerial jobs include chief executives; senior officials and legislators; administrative and commercial managers; production and specialized services managers; and hospitality, retail, and other services managers. Technical (e.g., analyst, engineer) jobs include professionals; technicians and associate professionals; skilled agricultural, forestry, and fishery workers; and craft and related trades workers. Administrative (e.g., secretary) jobs include general and keyboard clerks, customer services clerks, numerical and material recording clerks, and other clerical support workers. Customer-facing jobs include personal services workers, sales workers, personal care workers, and protective services workers. Manual jobs include plant and machine operators and assemblers; cleaners and helpers; agricultural, forestry, and fishery laborers; laborers in mining, construction, manufacturing and transport; food preparation assistants; street and related sales and services workers; refuse workers and other elementary workers.

Source: Based on a survey and structured interviews conducted with employers in the IT-BPO industry, n=32.

This impact is likely going to differ by occupation. For example, according to employers surveyed, manual and administrative jobs are likely to see the largest decreases due to employment as a result of 4IR technology adoption, with technical jobs seeing the largest increases (Figure 5). Box 1 describes the approach used to understand the potential employment impacts from 4IR.

This can have consequences for gender equity. Overall, more jobs held by males in the IT-BPO industry are likely to be affected (due to the greater number of males in the IT-BPO industry today). However, some occupations held predominantly by women may be particularly affected, such as those in administrative roles (Figure 6).

To understand what drives these results, it is important to first understand that technology does not automate jobs, but rather individual tasks or combinations of them. For example, in the case of IT-BPO, AI and big data applications will not replace a customer service agent or information technology (IT) professional, but it will replace the tasks of identifying customer solutions or directing the customer to the right department. A loss of employment only occurs if automation impacts such a high share of activities associated with a job that a worker is no longer essential.

Box 1: Estimating Employment Changes

This report employs an experimental approach to understanding the impact of Industry 4.0 (4IR) on employment. The core data sources used in this approach are the World Bank's Skills Measurement Program (STEP) survey for the Philippines, labor force survey (LFS) data, online job portal data and surveys of employers in the prioritized industries in the Philippines. The approach seeks to first understand how 4IR could impact the industry's growth trajectory and then how employment will change based on task shifts within occupations.

The growth trajectory of the industry is computed by looking at historic industry growth as a business-as-usual scenario and then modeling the impact of 4IR as a productivity shock that generates additional productivity growth. The assumption used for the estimates presented here is that adoption rates of 4IR increase to 50% until 2025 and from 2025 onwards, 4IR technology adoption grows to 100% by 2030. This approach is not meant to forecast the actual, or even a necessarily realistic, level of 4IR technology adoption by 2030. Rather, it should be considered a thought experiment to understand the largest possible impact 4IR can have on employment and skills gaps. Productivity shocks and adoption levels were obtained from the employer survey and cross-referenced with the broader literature, where available.

Estimating the changes in employment across different occupations relies on a detailed analysis of task profiles (Box 2). The analysis identifies the changes in the time spent on particular tasks between today and a future in which 4IR has been adopted. Combining this with the breakdown of employment by occupation for the industry as well as the productivity growth estimates above, the results show how different occupations may become more frequent in the industry. This part of the analysis mostly uses data from the STEP survey and Philippines' LFS.

Source: Asian Development Bank and AlphaBeta.

Figure 6: Modelled Displacement Effects of 4IR on Jobs Predominantly Held by Males vs. Females, 2018–2030

IT-BPO: Jobs

Technology displacement effects are likely to affect 12% more men than men in the IT-BPO industry in the Philippines

Displaced jobs held by males   (151,000)

Displaced jobs held by females   (135,000)

+12%

( ) = negative, 4IR = Industry 4.0 or Fourth Industrial Revolution, GDP = gross domestic product, IMF = International Monetary Fund, IT-BPO = information technology and business process outsourcing, LFS = labor force survey, PSA = Philippine Statistics Authority, STEP = Skills Measurement Program of the World Bank.

Note: Change in jobs based on accelerated adoption scenario of 4IR technologies.

Sources: Industry employment –PSA, LFS 2017 and ILO; GDP/Output –PSA and IMF Article IV; STEP survey data; Employer survey on impact of 4IR on the IT-BPO industry in the Philippines, n= 32+; Job portal data: jobs in the IT-BPO industry scraped from the job portal Bestjobs.ph from July to August 2019.

Job Task Implications

While understanding this dynamic of the interplay between tasks and technology is generally accepted in the literature, a limited amount of work has focused on emerging economies, including in Southeast Asia. Further, where such studies have been conducted, international data sources, often from developed countries like the United States, have been used in the analysis. This assumption is likely ill-founded as even identical occupations can vary significantly in their task profiles across countries.[11] As a result, this study adds to the literature by using local data sources rather than relying on international proxies. The research examines five types of tasks linked to jobs in the IT–BPO industry and how they could be impacted by 4IR:

(i)   **Routine physical.** These tasks involve repetitive and predictable physical work. For example, a factory worker assembling parts on a manufacturing line.

(ii) **Routine interpersonal.** These tasks involve predictable interactions with other people. For example, a call center worker reading a sales script.

(iii) **Nonroutine physical.** These tasks involve physical work that is not repetitive or predictable. For example, a mechanic diagnosing and repairing problems in a car engine.

(iv) **Nonroutine interpersonal.** These tasks involve complex or creative interactions with other people. For example, supervising others or making speeches or presentations.

(v)   **Analytical.** These are tasks that vary significantly and there is a strong thinking and/or analytical component. They predominantly involve computers or other technological equipment.

The research indicates an increase in the time spent on nonroutine tasks as well as analytical tasks. In fact, by 2030, workers in the industry could spend an additional 13.3% of their work week on such tasks and 13.3% less on routine physical and interpersonal tasks (Figure 7). The largest increase in time spent can be observed for analytical tasks. This is likely driven by the fact that technology will be able to automate responses to simple queries and undertaking basic processes, while humans will focus on more complex unique issues that will test their problem solving abilities. This could be call center agents responding to a customer query or unique problem, or IT operators dealing with a security breach. Box 2 describes the approach used to understand the potential impacts on 4IR on tasks in the workplace.

Skills Implications

These task shifts will potentially have significant implications for the aggregate skills required in the industry. The analysis considers 10 categories of skills:

(i)   **Critical thinking and adaptive learning.** Skills that allow the use of logic and reasoning to identify the strengths and weaknesses of alternative solutions, conclusions or approaches to problems. These skills also allow an understanding of the implications of new information for both current and future problem solving and decision-making.

(ii) **Written and verbal communication.** Ability to read, write, speak, and actively listen.

(iii) **Numeracy.** Ability to use mathematics and scientific rules and methods to solve problems.

(iv) **Complex problem solving.** Skills that help identify complex problems and review related information to develop and evaluate options and implement solutions.

(v)   **Management.** Skills that help allocate financial, material, personnel and time resources efficiently.

(vi) **Social.** Skills that help to work with people to achieve goals such as coordination, instructing, negotiation, persuasion, service orientation, and social perceptiveness or empathy.

---

[11]   W. Hardy et al. 2019. Technology, Skills, and Globalization: Explaining International Differences in Routine and Nonroutine Work Using Survey Data. *IBS working paper* 04/2019. https://ibs.org.pl/en/publications/technology-skills-and-globalization-explaining-international-differences-in-routine-and-nonroutine-work-using-survey-data/.

Figure 7: Shifts in Time Spent by Workers on Different Tasks at Work in the IT-BPO Industry, 2018–2030

IT-BPO: Tasks

4IR is likely to shift time spent on routine tasks to analytical and nonroutine tasks

Average share of weekly working hours spent on this task (%)

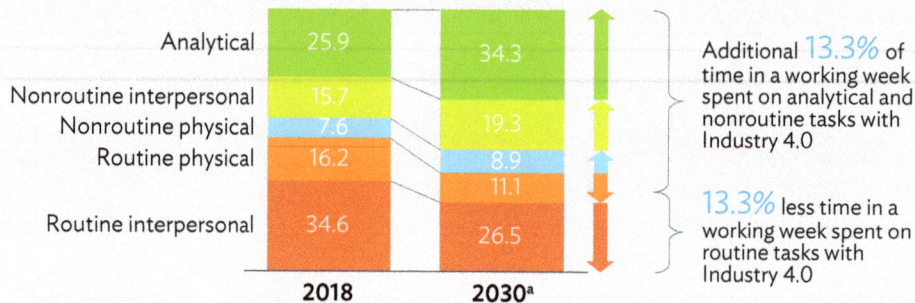

Task	2018	2030[a]
Analytical	25.9	34.3
Nonroutine interpersonal	15.7	19.3
Nonroutine physical	7.6	8.9
Routine physical	16.2	11.1
Routine interpersonal	34.6	26.5

Additional 13.3% of time in a working week spent on analytical and nonroutine tasks with Industry 4.0

13.3% less time in a working week spent on routine tasks with Industry 4.0

IT-BPO = information technology and business process outsourcing.

Note: Figures include rounding adjustments.

[a] Based on a "high adoption" scenario of Industry 4.0. The Appendix has more details.

Source: Asian Development Bank and AlphaBeta.

Box 2: Estimating Task Shifts

For this analysis, the report uses what the literature refers to as a task-based approach. It starts by identifying the employment breakdown of the industry according to occupations using LFS data. This provides an overview of the occupations in an industry and the relative employment by occupation for 43 occupations, aggregated into 5 major groups: managerial; technical (e.g., analyst, engineer); administrative (e.g., secretary); customer facing; manual (e.g., floor workers).

For each of the occupations in the industry a task profile was developed. A task profile gives a detailed description of how many hours on average per week the average worker in an occupation in this industry spends executing specific tasks. Based on the literature, the five different task groups listed above were identified.[a] To create individual task profiles for each of the occupations in the relevant industries, data collected by the World Bank's STEP program was used.[b] Questions from the survey were used to allocate time spent on task groups. First, the amount of time spent on routine vs. nonroutine tasks was determined, then each time allocation was further split between physical, interpersonal and analytical tasks. The result is a profile of the relative time, in terms of hours spent, of each the five task groups for each occupation in the industry.

To understand how these task profiles shift with 4IR technology adoption, estimates from the employer survey were used. Employers were asked to provide estimates of the change in aggregate time spent by task in their firm (i.e., change in the total time all workers in the firm spend on the set task collectively) due to 4IR technology adoption over the next 5 years. The fundamental assumption is that the adoption of 4IR technologies changes the task profile of an occupation through automation of certain tasks and time shifted to others. This results in new task profiles by occupation for 2030 where 100% of firms have adopted 4IR.

[a] Prospera and AlphaBeta Advisors. 2019. *Capturing Indonesia's Automation Potential*. https://www.alphabeta.com/wp-content/uploads/2019/08/capturing-indonesias-automation-potential.pdf.
[b] The STEP Skills Measurement Program. https://microdata.worldbank.org/index.php/catalog/step/about.

Source: Asian Development Bank and AlphaBeta.

(vii) **Evaluation, judgment, and decision-making.** Skills used to understand, monitor, conduct and improve analysis and socio-technical systems.

(viii) **Technical.** Skills used to design, set up, operate, maintain, and correct malfunctions involving the application of machines or technological systems.

(ix) **Computer literacy.** These are basic skills that allow workers to effectively use computers and digital applications in their jobs, such as email, word processing, searching the internet, data entry, etc.

(x) **Digital and information communication technology.** These are advanced skills that allow workers to work in inherently digital occupations and perform complex tasks in a digital environment, as well as operating and/or developing digital tools such as advanced spreadsheet functions, financial software, graphic design, statistical analysis, software programming or managing computer networks.[12]

Box 3 describes the approach used to understand the potential impacts on 4IR on the skills required by workers.

Box 3: Estimating Skills Changes

To compute current skills profiles for each occupation in the industry, data from the Skills Measurement Program (STEP) of the World Bank questionnaire's Module 6: Work Skills was used. Questions from this chapter were used to assess the importance of each skills category. A value from 0 to 3 (0 for a skill that is not used, to 3 for highly advanced skills being required) was assigned to skills based on survey responses to relevant questions. The score measured both the importance as well the competency level of the skill for each skills category.

Future skills profiles leveraged two sources of data: (i) Data on skill and education requirements from job profiles for occupations, obtained from online job portals; and (ii) information about changes in skills requirements from the employer survey.

The collected job postings were analyzed in detail and assigned an importance/skill competency score (from 0 to 3) for each of the 10 skills categories. They were also categorized according to the five job groups identified— managerial, administrative, technical, customer-facing, and manual.

In parallel, as a second estimate, survey data of employers was leveraged to understand which skills categories would gain in importance due to adoption of Industry 4.0 at an industry level. Based on the responses, percentage changes in the level-of-importance scores were calculated for the five job groups identified above. Applying these to the current skills profiles based on the STEP resulted in a second set of estimates for future skills profiles.

The future skills profiles used to estimate the skills gap were then computed as an average of the two estimates and the skills gap by occupation was identified by simply examining the differences in importance scores between current and future skills profiles.

Source: Asian Development Bank and AlphaBeta.

---

[12] The 10 skills categories and their definitions were chosen to align with the six skills groups provided by O*NET, which is one of the key databases for examining skills changes in the workforce. Some adjustments were made to the O*NET classifications to better align with the analysis. These included disaggregating O*NET's basic skills group into critical thinking and active learning, written and verbal communication, and numeracy. Basic digital and advanced digital/ICT skills were also broken out of O*NET's broader technical skills group due to their particular relevance for 4IR.

Figure 8: Potential Impact of 4IR on Shifts in the Importance of Different Skills in the IT-BPO Industry

IT-BPO: Skills

Evidence from job portal data does not support all employer sentiment about changes to skill importance due to industry 4.0 adoption

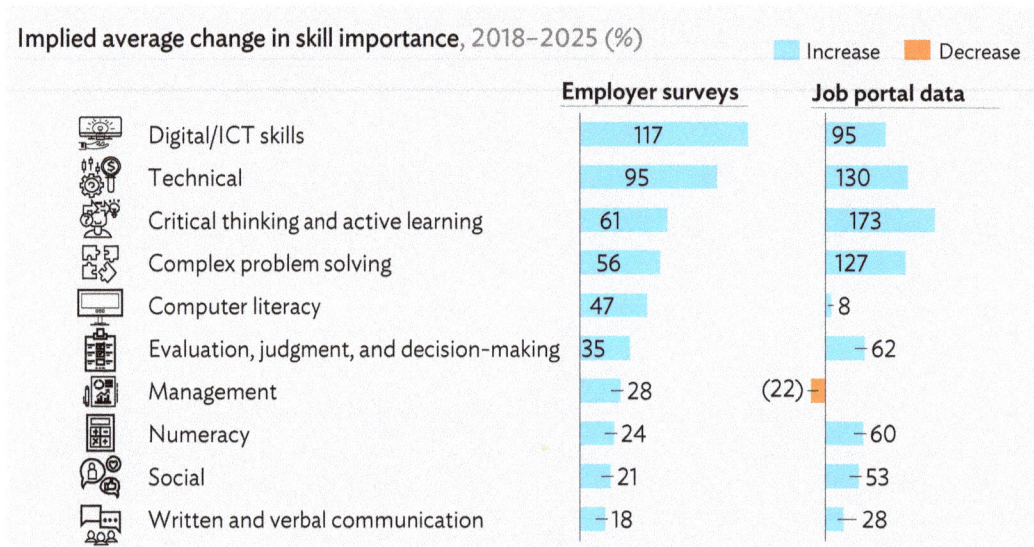

Implied average change in skill importance, 2018–2025 (%)   ■ Increase ■ Decrease

Skill	Employer surveys	Job portal data
Digital/ICT skills	117	95
Technical	95	130
Critical thinking and active learning	61	173
Complex problem solving	56	127
Computer literacy	47	8
Evaluation, judgment, and decision-making	35	62
Management	-28	(22)
Numeracy	-24	60
Social	-21	53
Written and verbal communication	-18	28

( ) = negative, 4IR = Industry 4.0 or Fourth Industrial Revolution, GDP = gross domestic product, ICT = information and communication technology, IMF = International Monetary Fund, IT-BPO = information technology and business process outsourcing, LFS = labor force survey, PSA = Philippine Statistics Authority, STEP = Skills Measurement Program of the World Bank.

Sources: Industry employment –PSA, LFS 2017 and ILO; GDP/Output –PSA and IMF Article IV; STEP survey data; Employer survey on impact of 4IR on the IT-BPO industry in the Philippines, n= 32+; Job portal data: jobs in the IT-BPO industry scraped from the job portal Bestjobs.ph from July to August 2019.

Based on the skills categories, unique, current, and future (i.e., after 4IR technology adoption) skills profiles for occupations in the industry were created based on data from the World Bank's STEP survey, job portals, and inputs from the employer survey. These profiles were then compared to understand the skills gap created by 4IR technology adoption.

The analysis highlights some significant changes in the skills requirements in the IT-BPO industry:

(i)   **Change in skills demand.** This study reviewed both employer survey data and job portal data to understand changes in the importance of skills linked to 4IR. Interestingly, while employers perceived digital/ICT skills to be the fastest growing skills categories, the job portal data reflected these to be critical thinking and active learning and complex problem solving skills (Figure 8). Further, job portal data also pointed toward numeracy and social skills being more important than employers are anticipating. Interestingly, job portal data is predicting an absolute, not just relative, decline in the importance of management skills. These insights align with the insights from the July 2019 stakeholder consultation workshop held in Manila, where industry participants emphasized that the skills required in the IT-BPO industry previously, such as communication (e.g., language skills) and social skills (e.g., service

Figure 9: Impact of 4IR on the Importance of Different Skills in the IT-BPO Industry, 2018–2030

IT-BPO: Skills

Evaluation, judgment and decision-making skills will likely be crucial as 4IR technology deals with routine processes

- Skills of increasing relative importance from 2018–2030
- Skills with decreasing relative importance from 2018–2030
- Skills with no change in relative importance

Importance ranking	2018	2030
1	Written and verbal communication	Evaluation, judgment, and decision-making
2	Management	Numeracy
3	Numeracy	Written and verbal communication
4	Social	Social
5	Evaluation, judgment, and decision-making	Computer literacy
6	Computer literacy	Critical thinking and active learning
7	Complex problem solving	Complex problem solving
8	Critical thinking and active learning	Management
9	Technical	Technical
10	Digital/ICT skills	Digital/ICT skills

( ) = negative, 4IR = Industry 4.0 or Fourth Industrial Revolution, GDP = gross domestic product, ICT = information and communication technology, IMF = International Monetary Fund, IT-BPO = information technology and business process outsourcing, LFS = labor force survey, PSA = Philippine Statistics Authority, STEP = Skills Measurement Program of the World Bank.

Sources: Industry employment –PSA, LFS 2017 and ILO; GDP/Output –PSA and IMF Article IV; STEP survey data; Employer survey on impact of 4IR on the IT-BPO industry in the Philippines, n= 32+; Job portal data: jobs in the IT-BPO industry scraped from the job portal Bestjobs.ph from July to August 2019.

orientation) will no longer be sufficient for employment. However, skilled workers with specialized training and knowledge, and the ability to handle complex situations requiring advanced levels of perception and critical thinking, will likely thrive in a 4IR environment (footnote 5).

(ii) **Overall skills importance.** Evaluation, judgment, and decision-making and numeracy skills are likely to see their relative importance increase by 2030, replacing communication skills as the most important skills in the industry, although these will continue to be important, potentially ranking third in 2030 (Figure 9). While computer literacy and critical thinking and active learning will see relative increases in importance, management skills are forecast to see the greatest relative decline in skills importance. One reason could be that 4IR technologies will mean that less management is needed, or more detailed data, business intelligence, and AI assistance will allow for easier management of people and resources, especially from remote locations. For IT-BPO industries, this could mean that managers may not have to be physically located with their teams.

(iii) **Changes in level of skills.** Overall, the industry will require significant upskilling as the demand for more intermediate and advanced skills is likely to increase (Figure 10). Given the broad coverage of BPO, the industry is expected to see significant increases in demand for intermediate and advanced skills across all skills categories.

Figure 10: Impact of 4IR on the Level of Skills Required in IT-BPO, 2018–2030

IT-BPO: Skills

4IR adoption will require large increases in intermediate critical thinking and communication, as well as advance social and evaluation skills

Skills	Absolute Change in Percentage of Workers Requiring Skill by Level, 2018–2030 (%)		
	Basic	Intermediate	Advanced
Critical thinking and active learning	(84.2)	38.4	50.0
Written and verbal communication	(2.3)	(75.1)	77.4
Numeracy	(0.3)	(99.4)	99.7
Complex problem solving	(77.9)	41.0	41.1
Management	0.0	(12.3)	12.3
Social	(22.9)	(53.3)	76.3
Evaluation, judgment, and decision-making	(3.8)	(96.2)	100.0
Technical	(79.0)	74.1	19.6
Computer literacy	(28.6)	(43.8)	72.4
Digital/ICT skills	(39.5)	84.7	15.1

Legend:
- >50%
- >10%
- ≤10%; ≥-10%
- < -10%
- < -50%

( ) = negative, 4IR = Industry 4.0 or Fourth Industrial Revolution, GDP = gross domestic product, ICT = information and communication technology, IMF = International Monetary Fund, IT-BPO = information technology and business process outsourcing, LFS = labor force survey, PSA = Philippine Statistics Authority, STEP = Skills Measurement Program of the World Bank.

Sources: Industry employment –PSA, LFS 2017 and ILO; GDP/Output –PSA and IMF Article IV; STEP survey data; Employer survey on impact of 4IR on the IT-BPO industry in the Philippines, n= 32+; Job portal data: jobs in the IT-BPO industry scraped from the job portal Bestjobs.ph from July to August 2019.

Box 4: Comparison of Insights in the IT-BPO Industry versus Past Research

According to the International Labour Organization (ILO), 89% of Filipino business process outsourcing workers are at high risk of automation. While the research presented in this report predicts large disruption due to potential job losses of 24%, the key message is a lot more encouraging.[a] Due to the income effect created by productivity gains from 4IR technology, the net effect on employment is likely to be positive.

The insights on skills shifts are also largely consistent with the ILO analysis. The ILO found technical knowledge, here defined as technical skills, to be the most critical skill according to Philippine enterprises across the entire economy. This was followed by teamwork and strategic thinking, classified here under social and critical thinking.

IT-BPO = information technology and business process outsourcing.

[a] Bureau for Employers' Activities and International Labour Organization. 2017. *ASEAN in Transformation —How Technology is Changing Jobs and Enterprises—The Philippines Country Brief.* https://www.ilo.org/actemp/publications/WCMS_579667/lang--en/index.htm.

Source: Asian Development Bank and AlphaBeta.

Skills Supply Trends

Figure 11 shows the breakdown of the additional demand for training that will be required by workers in the IT-BPO industry under 4IR technology adoption. This reflects the amount of training required to bring the IT-BPO workforce in the Philippines from the skills required in 2018 to the level of skills required by 2030, driven only by 4IR technology adoption. Overall, there will need to be 14.2 million additional person trainings by 2030. The majority of the training requirements will likely come from on-the-job training, with the rest evenly split between short professional training and longer formal training.

In the IT-BPO industry, the number of workers who will remain in employment (i.e., are unlikely to lose their jobs) but require upskilling far outweighs the number of workers who will lose their jobs or even new workers entering the industry due to the net employment effect. For these workers, on-the-job training might be the most suitable training type. This also highlights the need for the formal technical and vocational education and training (TVET) sector to shift its focus from long-term training and education to also being actively involved in shorter professional and even on-the-job training.

Figure 11: Additional Person Trainings Required to Meet Skills Demand Driven by 4IR Adoption in IT-BPO by Training Channel in 2030

IT-BPO: Training

59% of the additional demand for training driven by 4IR adoption will likely need to be met by "on-the-job" training

Millions of person trainings required by channel

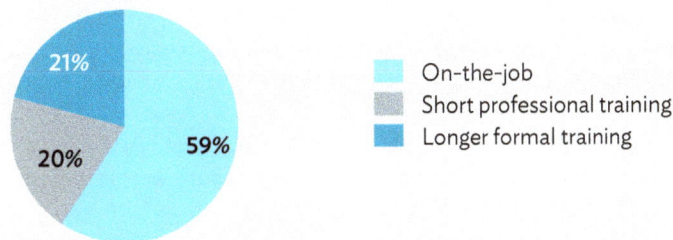

- On-the-job
- Short professional training
- Longer formal training

4IR = Industry 4.0 or Fourth Industrial Revolution, GDP = gross domestic product, ICT = information and communication technology, IMF = International Monetary Fund, IT-BPO = information technology and business process outsourcing, LFS = labor force survey, PSA = Philippine Statistics Authority, STEP = Skills Measurement Program of the World Bank.

Notes:
1. Figures include rounding adjustments.
2. One person training refers to training one worker, in one skill from the level required by his occupation's skill profile in 2018 to the relevant level given by the skills profile in 2030.
3. "On-the-job" training refers to training conducted during day to day such as senior staff instructing junior staff or running internal seminars; "Short professional" training refers to short (1 day to 6 months) courses conducted by professional internal or external instructors (e.g., weekend seminars, boot-camps); and "Longer formal" trainings refer to trainings longer than 6 months for which workers would likely have to take leave from their jobs, these include returning into formal education such as obtaining a degree.

Sources: Industry employment –PSA, LFS 2017 and ILO; GDP/Output –PSA and IMF Article IV; STEP survey data; Employer survey on impact of 4IR on the IT-BPO industry in the Philippines, n= 32+; Job portal data: jobs in the IT-BPO industry scraped from the job portal Bestjobs.ph from July to August 2019.

Box 5: Estimating Training Requirements

Skills supply, or in other words, training requirements, can be quantified in person training. One person training refers to training one worker in one skill from the skill level specified by the person's occupational skill profile in the industry in 2018 to the required level under 4IR technology adoption. Hence, the training required to shift a worker from the worker's current skills profile to the future skills profile would require one person training per skill that needs improvement between 2018 and 2030.

To understand the type of training needed, and in particular, the length of training, two factors need to be considered: (i) understanding the level of skills improvement needed, and (ii) understanding the access to different training channels of different workers.

Individuals' training needs are going to differ if they need to improve their skills from basic to intermediate levels, intermediate to advanced levels, or even basic to advanced levels. For example, a worker who only requires basic technical skills today but requires advanced technical skills in 2030 under 4IR technology adoption will likely require more training than another worker who only has to improve skills from intermediate to advanced levels. The same applies to workers who do not need a particular skill today but will require it in 2030, whether basic, intermediate, or advanced.

Apart from the length of training required to obtain a certain level of skills, access to different channels of training may not be the same for all workers in the industry. For example, workers who lose their jobs may not be able to receive on-the-job training but instead require formal training prior to being able to find new employment. Similarly, for future generations of workers (i.e., students currently in formal education or training) it may make more sense to embed skills training in their formal curriculum rather than waiting to train them on the job. Three categories of workers impacted by 4IR were identified based on the skills demand analysis:

(i)   **Workers in need of reskilling.** These workers will likely lose their current jobs due to automation, meaning they need to receive training that makes them employable in new jobs created.
(ii) **Workers in need of upskilling.** These workers will likely remain in their occupations, but the adoption of 4IR technologies means they will have to acquire new skills as well as improve existing skills to upgrade to their occupation's future skills profile.
(iii) **Future workers.** These will be additional workers required to fill the jobs generated from growing demand. Hence, they are workers who have not previously work in the industry, but they could either join as new graduates or professional hires from other industries.

The distinction of the type of worker is important, as different workers have access to different types of training. For example, while future workers are likely to receive some of their skills training in their formal education, returning to formal education is an unlikely option for workers in need of upskilling, who will continue to be employed during their training.

Source: Asian Development Bank and AlphaBeta.

To better understand the Philippines' training and education sector, two surveys were conducted. As part of the survey of employers in the IT-BPO industry, respondents were asked to comment on their ability to attract good candidates for jobs as well as their current engagement in training. A separate training institution survey was commissioned, the results of which are discussed in Chapter 2.

Figure 12: Employer Sentiment Toward Graduates in the Past 24 Months in IT–BPO

IT-BPO: Training

On average, employers seem to be satisfied with the quantity and quality of graduates for the IT-BPO industry

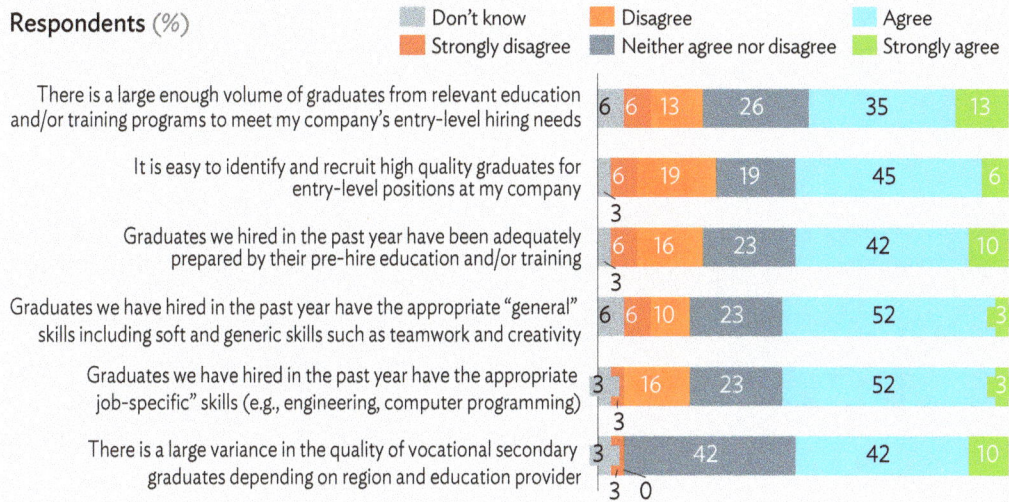

Respondents (%)

Legend: Don't know | Disagree | Agree | Strongly disagree | Neither agree nor disagree | Strongly agree

Statement	Don't know	Strongly disagree	Disagree	Neither agree nor disagree	Agree	Strongly agree
There is a large enough volume of graduates from relevant education and/or training programs to meet my company's entry-level hiring needs	6	6	13	26	35	13
It is easy to identify and recruit high quality graduates for entry-level positions at my company	6	3	19	19	45	6
Graduates we hired in the past year have been adequately prepared by their pre-hire education and/or training	6	3	16	23	42	10
Graduates we have hired in the past year have the appropriate "general" skills including soft and generic skills such as teamwork and creativity	6	6	10	23	52	3
Graduates we have hired in the past year have the appropriate job-specific" skills (e.g., engineering, computer programming)	3	3	16	23	52	3
There is a large variance in the quality of vocational secondary graduates depending on region and education provider	3	3	0	42	42	10

4IR = Industry 4.0 or Fourth Industrial Revolution, IT–BPO = information technology and business process outsourcing.
Source: Employer survey on impact of 4IR on the IT–BPO industry in the Philippines, n=31.

The survey results of employers in the IT-BPO industry reveal that less than half of employers feel that there is a sufficient volume of graduates from relevant education and training programs to meet their company's entry-level hiring needs, and only 51% of employers find it easy to identify high quality graduates (Figure 12). More than 50% agreed (or strongly agreed) that graduates hired in the last year had been adequately prepared by their previous education or training. Interestingly, few employers strongly agreed with many of these statements, raising the question as to whether these figures might appear more positive than they actually are. This is echoed by industry interviews that highlighted significant challenges in finding suitably qualified graduates.

Electronics Manufacturing Industry

Results from the electronics manufacturing industry are very consistent with those of the IT-BPO industry. Similar to the IT-BPO industry, the overall impact on employment is estimated to be net positive, driven by strong productivity gains associated with 4IR. The automation impact is likely to be felt mainly on manual jobs, which tend to be held mostly by women; this could create adverse effects on inclusiveness in the industry.

As opposed to the IT-BPO industry, technical skills are likely to be of far greater importance in the electronics manufacturing industry under 4IR technology adoption, requiring far more workers with advanced skills. As with the IT-BPO industry, much of the additional training required will have to take place on the job, meaning the TVET sector may have to consider adjusting its offering to these workers.

Another issue could be that the positive net effect on employment could put further pressure on the supply of workers as 33% of employers feel there is not a sufficiently large volume of graduates.

To achieve this, a significant amount of additional training will be required between 2018 and 2030. Overall, there will need to be 7.5 million additional person trainings by 2030. As well as on-the-job training, there will be a significant amount of longer formal training required to bridge some significant skills gaps in the industry.

Relevance to Industry 4.0

4IR technology adoption in the electronics manufacturing industry is being driven by several macroeconomic trends. These include:

(i)   **Global supply chains.** Electronics manufacturers have locations all over the world and 4IR technologies can enable seamless connectivity between these locations to enable operations to shift in response to production or demand fluctuations.

(ii)   **More complex designs and shorter deliver times.** 4IR technologies are critical to responding to the increasingly complex designs of electronic components and demands for shorter turnaround times.

(iii) **Greater focus on reliability of production systems.** As facilities increasingly move toward 24-hour production, equipment reliability becomes even more critical. 4IR-enabled plants have robust monitoring systems to identify potential maintenance issues before they cause downtime.

There are various 4IR technologies of relevance to the electronics manufacturing industry, ranging from digital technologies enabling smart factories through to additive manufacturing enabling mass customization of products. McKinsey & Company has estimated that productivity could be increase by 30%–50% through adopting relevant technologies in the electronics manufacturing industry in ASEAN.[13]

Some key technologies include:

(i)   **Internet of Things.** IoT refers to networks of sensors and actuators embedded in machines and other physical objects that connect with one another and the Internet. It has a wide range of applications, including data collection, monitoring, decision-making, and process optimization.[14] Radio frequency identification tags on containers can track products as they move from the factory to stores, allowing companies to avoid stockouts and losses.

(ii)   **Artificial intelligence and big data.** Big data refers to the ability to analyze extremely large volumes of data, extract insights, and act on them closer to real time. This has a range of benefits in the electronics manufacturing industry, including being able to use predictive analytics to fine-tune production volumes and processes, for better supply chain management, and for greater insights into customer segments.

(iii) **Industry robotics.** The electronics manufacturing industry is forecast to soon surpass the automotive industry as the most important source of demand for industry robots. In 2017, the share of sales to the automotive industry was only 1 percentage point more than the share

13   I. Arbulu et al. 2018. *Industry 4.0: Reinvigorating ASEAN Manufacturing for the Future.* McKinsey & Company. 8 February. https://www. mckinsey.com/business-functions/operations/our-insights/industry-4-0-reinvigorating-asean-manufacturing-for-the-future.

14   J. Woetzl et al. 2014. *Southeast Asia at the Crossroads: Three Paths to Prosperity. McKinsey Global Institute.* November. https:// www.mckinsey.com/~/media/McKinsey/Featured%20Insights/Asia%20Pacific/Three%20paths%20to%20sustained%20 economic%20growth%20in%20Southeast%20Asia/MGI%20SE%20Asia_Executive%20summary_November%202014.ashx.

of sales to electronics manufacturing firms. Based on current growth rates, by 2021, the sales of robots to the electronics manufacturing industry will be higher than that of the automotive industry.[15] Industrial robots can significantly improve productivity in electronics manufacturing and there has been increased investment in robots in this industry in response to drivers such as wage costs, the need for full-time (24/7) production, and high levels of staff turnover. Another key driver has been the expanding range of tasks that robots can perform, particularly in the assembly of electronic components and equipment. Robots can now be used across the entire production cycle—from cutting metal housings to assembling miniature components on boards, applying sealants and adhesives, buffing and polishing surfaces, performing quality inspections, and packing and palletizing finished products; developments in sensors and power-force limiting technologies (that ensure the robot slows down or stops if it comes into contact with a worker) mean that robots can now share workspaces with human employees (footnote 15).

(iv) **Additive manufacturing (3D printing).** This describes the technologies that build 3D objects by adding layer upon layer of material. There is a range of potential benefits for electronic products, such as ability to handle complex, low-volume components where rapid turnaround is critical.[16]

In the electronics manufacturing industry, 56% of the businesses surveyed say they have a good understanding of what technologies are relevant to their industry (Figure 13), 49% agree that they are already adopting 4IR technologies to some degree and 10% more are planning to adopt 4IR over the next 5 years. However, there are some concerns about the cost of 4IR technology adoption (as in the IT-BPO industry), since 53% of employers see the cost as a significant barrier.

The likely reason that employers in electronics manufacturing are eager to adopt 4IR is a general belief in high productivity returns. According to the survey data, 50% of employers in the industry in the Philippines believe that they can expect productivity improvements from 4IR to fall by 25%–100% over the next 5 years (Figure 14). This is in line with the research published by McKinsey & Company mentioned previously.[17]

Skills Demand Analysis

Employment Implications

The results from the electronics manufacturing industry are very similar to those from the IT-BPO industry (Figure 15). Approximately 24% of jobs are at risk of being lost or automated. Also, the overall effect on employment is likely to be net positive, with an increase of around 10% of current jobs predicted by 2030 due to 4IR.

There are a number of drivers behind these similarities. The aggregate task profiles of the industries are very similar, despite the fact that IT-BPO is a service-oriented industry and electronics manufacturing is a manufacturing industry. Both industries are expected to see similar shifts in time spent on different tasks, according to employers surveyed and international estimates.

---

[15] International Federation of Robotics. 2018. Automation boom in electrical /electronics industry drives 30% increase in sales of industrial robots. 7 November. https://ifr.org/post/automation-boom-in-electrical-electronics-industry-drives-30-increase-in-sa.

[16] N. Sharp. 2019. Is Additive Manufacturing the Right Choice for Your Electronic Assembly? 7 November. https://blog. jjsmanufacturing.com/additive-manufacturing-electronic-assembly.

[17] McKinsey & Company. 2018. *Industry 4.0: Reinvigorating ASEAN Manufacturing for the Future.* https://www.mckinsey.com/business-functions/operations/our-insights/industry-4-0-reinvigorating-asean-manufacturing-for-the-future.

Figure 13: Sentiments Toward 4IR in the Electronics Industry

Electronics: 4IR readiness

Companies in the electronics industry claim to have a good understanding of 4IR and high adoption

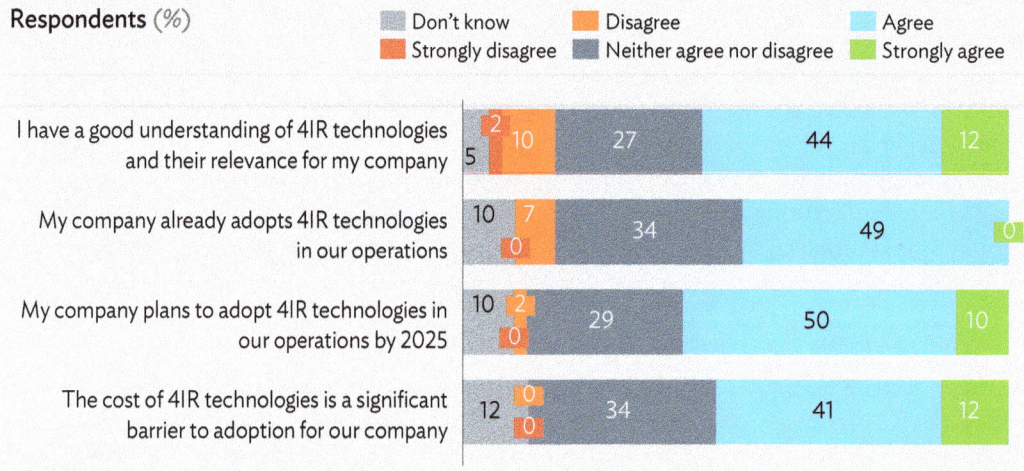

Respondents (%)

Legend: Don't know | Disagree | Agree | Strongly disagree | Neither agree nor disagree | Strongly agree

Statement	Strongly disagree	Disagree	Neither	Agree	Strongly agree
I have a good understanding of 4IR technologies and their relevance for my company	5 / 2	10	27	44	12
My company already adopts 4IR technologies in our operations	10 / 0	7	34	49	0
My company plans to adopt 4IR technologies in our operations by 2025	10 / 0	2	29	50	10
The cost of 4IR technologies is a significant barrier to adoption for our company	12 / 0	0	34	41	12

4IR = Industry 4.0 or Fourth Industrial Revolution.

Source: Employer survey on impact of 4IR on the electronics industry in the Philippines, n=41.

Figure 14: Expected Productivity Improvement Due to 4IR Technologies in 5 Years

Electronics: Jobs

Over 50% of employers in the electronics industry expect a productivity increase between 25%–100% from 4IR technologies over the next 5 years

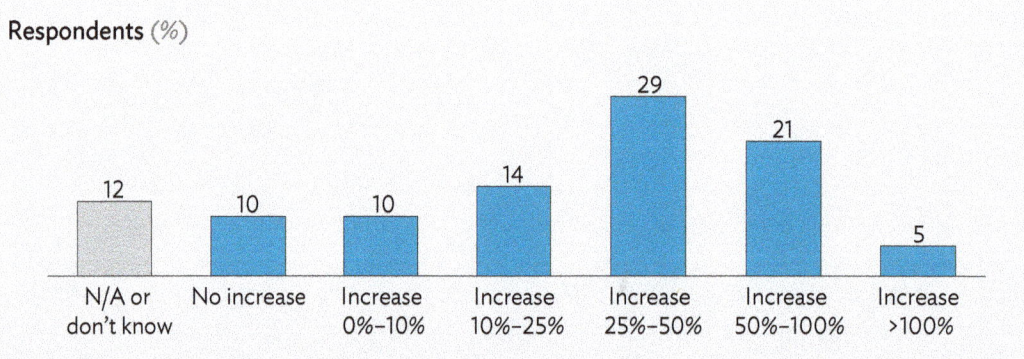

Respondents (%)

Category	%
N/A or don't know	12
No increase	10
Increase 0%–10%	10
Increase 10%–25%	14
Increase 25%–50%	29
Increase 50%–100%	21
Increase >100%	5

4IR = Industry 4.0 or Fourth Industrial Revolution, N/A = not applicable.

Source: Employer survey on impact of 4IR on the electronics industry in the Philippines, n=41.

Figure 15: Modelled Impact of 4IR on Number of Jobs in Electronics Manufacturing, 2018–2030

Electronics: Jobs

The overall impact of 4IR on jobs is likely an increase by over 30% of today's employment as displacement is potentially offset by income effects

Displacement and income effects of 4IR on jobs, 2018–2030 (%)

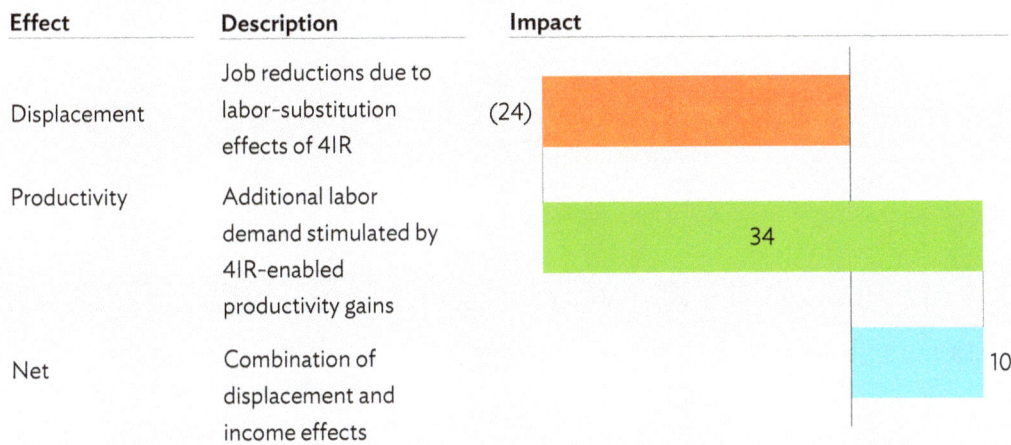

Effect	Description	Impact
Displacement	Job reductions due to labor-substitution effects of 4IR	(24)
Productivity	Additional labor demand stimulated by 4IR-enabled productivity gains	34
Net	Combination of displacement and income effects	10

( ) = negative, 4IR = Industry 4.0 or Fourth Industrial Revolution, GDP = gross domestic product, IMF = International Monetary Fund, LFS = labor force survey, PSA = Philippine Statistics Authority, STEP = Skills Measurement Program of World Bank.

Note: Change in jobs based on accelerated adoption scenario of 4IR technologies.

Sources: Industry employment – PSA, LFS 2017 and ILO; GDP/Output – PSA, McKinsey and IMF Article IV; STEP survey data; Employer survey on impact of 4IR on the electronics industry in the Philippines, n= 41+; Job portal data: jobs in the Electronics industry scraped from the job portal Bestjobs.ph from July to August 2019.

Given the same challenges to realizing the positive income effect, as in the IT-BPO industry, the Philippines' approach to skills development will be critical in realizing a positive labor market outcome related to 4IR in this industry. Only with suitable skills in the local workforce will workers who lose their jobs be able to move seamlessly into newly created jobs.

Another potential concern is that while industry employment is fairly evenly split between men and women, the job losses could particularly impact female workers: 73% of jobs lost will be manual jobs, mainly assemblers, which are more likely to be jobs held by women. Administrative jobs, which employers agree to be at the highest risk of losses due to 4IR, are also heavily female-dominated. Of all jobs lost, 55% are likely to be occupied by women, over 96,000 (Figure 16).

Job Task Implications

On aggregate, a larger task shift is expected for the electronics manufacturing industry as opposed to the IT-BPO industry. According to the results, workers in the industry could spend an additional 15.8% of their work week on interpersonal and other nonroutine tasks and 15.8% less on routine physical and routine interpersonal tasks (Figure 17). The share of working time spent on routine physical tasks is

Figure 16: Modelled Displacement Effects of 4IR on Jobs Predominantly Held by Males versus Females in Electronics Manufacturing, 2018–2030

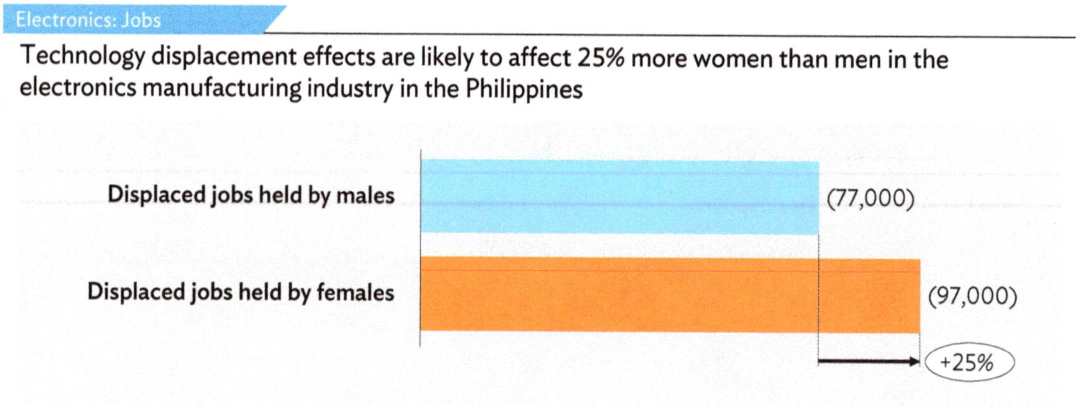

Electronics: Jobs

Technology displacement effects are likely to affect 25% more women than men in the electronics manufacturing industry in the Philippines

Displaced jobs held by males (77,000)

Displaced jobs held by females (97,000) +25%

4IR = Industry 4.0 or Fourth Industrial Revolution, GDP = gross domestic product, IMF = International Monetary Fund, LFS = labor force survey, PSA = Philippine Statistics Authority, STEP = Skills Measurement Program of World Bank.

Note: Change in jobs based on accelerated adoption scenario of 4IR technologies.

Sources: Industry employment – PSA, LFS 2017 and ILO; GDP/Output – PSA, McKinsey and IMF Article IV; STEP survey data; Employer survey on impact of 4IR on the electronics industry in the Philippines, n= 41+; Job portal data: jobs in the Electronics industry scraped from the job portal Bestjobs.ph from July to August 2019.

Figure 17: Shifts in Time Spent by Workers on Different Types of Tasks in Electronics Manufacturing, 2018–2030

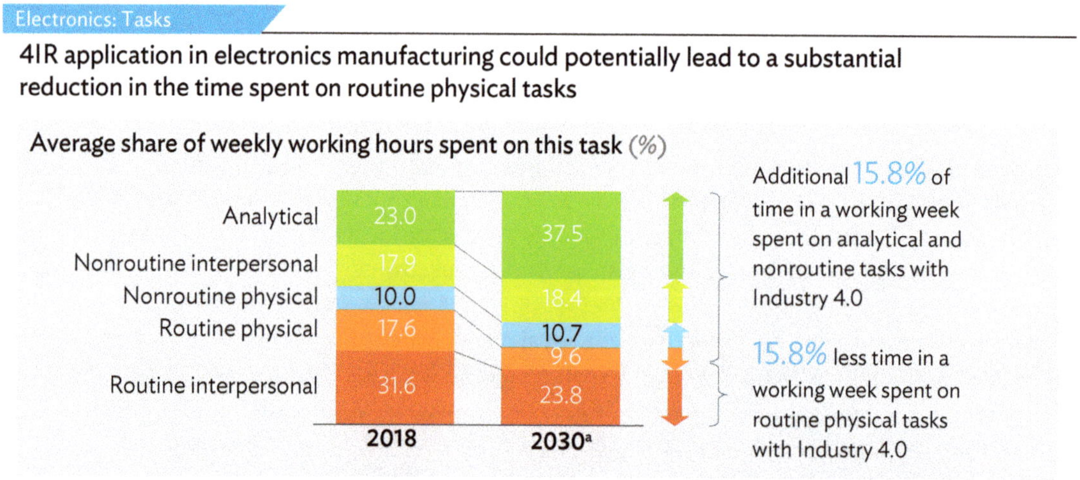

Electronics: Tasks

4IR application in electronics manufacturing could potentially lead to a substantial reduction in the time spent on routine physical tasks

Average share of weekly working hours spent on this task (%)

Task	2018	2030[a]
Analytical	23.0	37.5
Nonroutine interpersonal	17.9	18.4
Nonroutine physical	10.0	10.7
Routine physical	17.6	9.6
Routine interpersonal	31.6	23.8

Additional 15.8% of time in a working week spent on analytical and nonroutine tasks with Industry 4.0

15.8% less time in a working week spent on routine physical tasks with Industry 4.0

4IR = Industry 4.0 or Fourth Industrial Revolution, GDP = gross domestic product, IMF = International Monetary Fund, LFS = labor force survey, PSA = Philippine Statistics Authority, STEP = Skills Measurement Program of World Bank.

Note: Figures include rounding adjustments.

[a] Based on a "high adoption" scenario of 4IR.

Sources: Industry employment – PSA, LFS 2017 and ILO; GDP/Output – PSA, McKinsey and IMF Article IV; STEP survey data; Employer survey on impact of 4IR on the electronics industry in the Philippines, n= 41+; Job portal data: jobs in the Electronics industry scraped from the job portal Bestjobs.ph from July to August 2019.

Figure 18: Employers' Expected Impact of 4IR on Working Time Spent on Different Tasks in Electronics Manufacturing, 2018–2030

Electronics: Tasks

The majority of employers believes that the time spent on nonroutine tasks will be unaffected by 4IR adoption

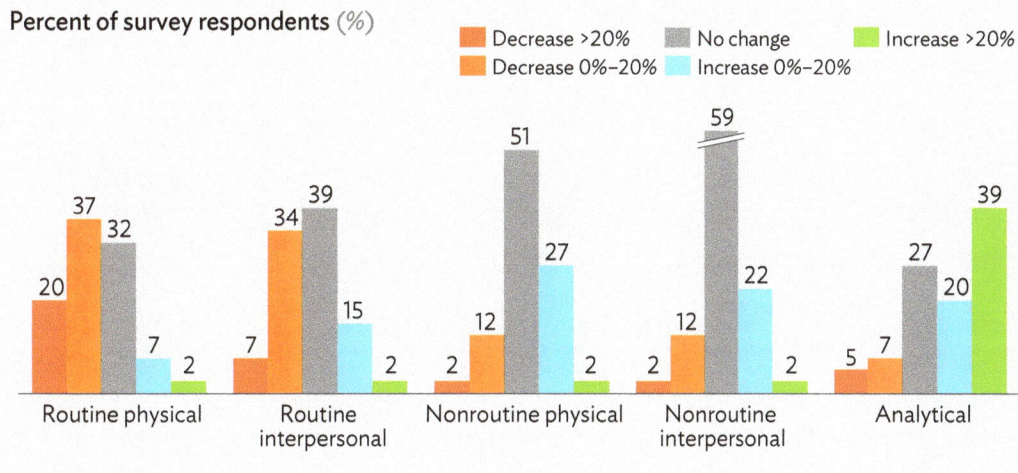

Percent of survey respondents (%)

Legend:
- Decrease >20%
- Decrease 0%–20%
- No change
- Increase 0%–20%
- Increase >20%

Routine physical: 20, 37, 32, 7, 2
Routine interpersonal: 7, 34, 39, 15, 2
Nonroutine physical: 2, 12, 51, 27, 2
Nonroutine interpersonal: 2, 12, 59, 22, 2
Analytical: 5, 7, 27, 20, 39

4IR = Industry 4.0 or Fourth Industrial Revolution.

Note: Answers in above chart do not sum to 100% as figure for the response "Don't know" was not included.

Source: Employer survey on impact of 4IR on the electronics industry in the Philippines, n=41.

predicted to almost halve, which, given the large share of assembly line work in this industry (with a high potential to be automated by 4IR technologies) is perhaps not surprising. Routine interpersonal tasks could decrease by a third.

This is consistent with insights from the employer survey, with 57% of employers expecting routine physical tasks to decline in the future (Figure 18).

Skills Implications

The analysis highlights some significant changes in the skills requirements in the industry:

(i)   **Change in skills demand.** According to employers, digital/ICT skills, technical, critical thinking and complex problem solving skills will see the biggest increases in importance over the next 5 years. This resonates with data extracted from job portals online. However, the online job portal data showed a slightly greater change in importance for technical skills (Figure 19). By contrast to the IT-BPO industry, the data also predicts that numeracy, social, and communication skills are unlikely to see major increases in importance, partially because they are already very important.

Figure 19: Potential Impact of 4IR on Shifts in the Importance of Different Skills in Electronics Manufacturing

Electronics: Skills

Numeracy, social, and "written and communication" skills are unlikely to see large increases in importance

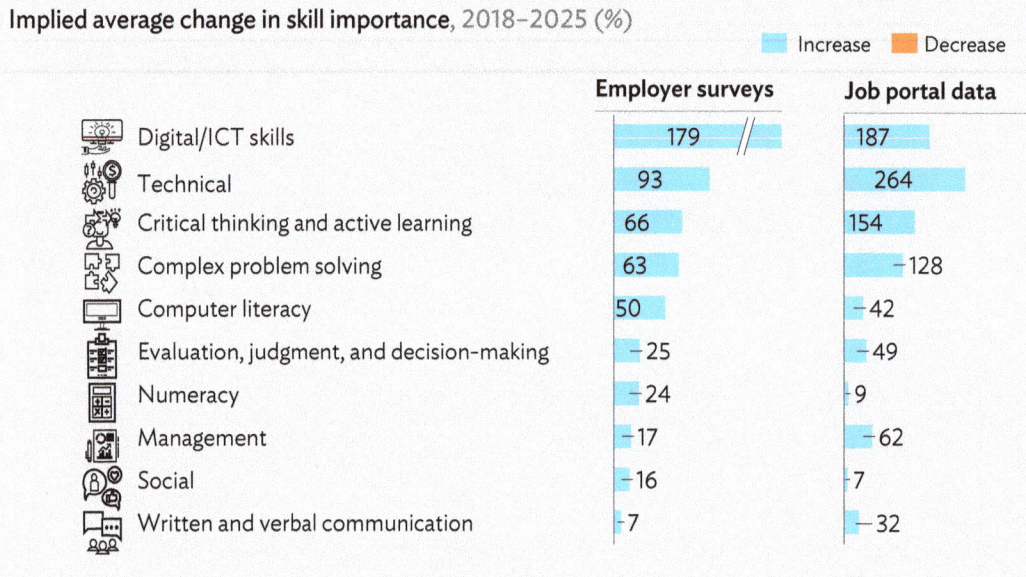

Implied average change in skill importance, 2018–2025 (%)

■ Increase   ■ Decrease

	Employer surveys	Job portal data
Digital/ICT skills	179	187
Technical	93	264
Critical thinking and active learning	66	154
Complex problem solving	63	−128
Computer literacy	50	−42
Evaluation, judgment, and decision-making	−25	−49
Numeracy	−24	9
Management	−17	−62
Social	−16	7
Written and verbal communication	7	−32

4IR = Industry 4.0 or Fourth Industrial Revolution, GDP = gross domestic product, ICT = information and communication technology, IMF = International Monetary Fund, LFS = labor force survey, PSA = Philippine Statistics Authority, STEP = Skills Measurement Program of World Bank.

Sources: Industry employment – PSA, LFS 2017 and ILO; GDP/Output – PSA, McKinsey and IMF Article IV; STEP survey data; Employer survey on impact of 4IR on the electronics industry in the Philippines, n= 41+; Job portal data: jobs in the Electronics industry scraped from the job portal Bestjobs.ph from July to August 2019.

(ii) **Overall skills importance.** Numeracy; evaluation, judgment, and decision-making; technical; and computer literacy skills will likely be the most important skills by 2030 due to 4IR technology adoption (Figure 20). Technical and computer literacy skills in particular are estimated to see a large increase in relative importance whereas communication, management, and social skills are forecast to see the greatest relative decline in importance.

(iii) **Changes in level of skills.** Overall, the industry will require significant upskilling as the demand for more intermediate and advanced skills is likely to increase (Figure 21). With the exception of social skills, all skill categories will require significant increases in workers with intermediate or advanced skills.

Figure 20: Impact of 4IR on the Importance of Different Skills in Electronics Manufacturing, 2018–2030

Electronics: Skills

The list of the most important skills in the electronics manufacturing industry is likely to change due to 4IR technology adoption

■ Skills of increasing relative importance from 2018–2030   ■ Skills with decreasing relative importance from 2018–2030

Importance ranking	2018	2030
1	Written and verbal communication	Numeracy
2	Management	Evaluation, judgment, and decision-making
3	Evaluation, judgment, and decision-making	Technical
4	Numeracy	Computer literacy
5	Social	Written and verbal communication
6	Computer literacy	Management
7	Complex problem solving	Complex problem solving
8	Critical thinking and active learning	Critical thinking and active learning
9	Technical	Social
10	Digital/ICT skills	Digital/ICT skills

4IR = Industry 4.0 or Fourth Industrial Revolution, GDP = gross domestic product, ICT= information and communication technology, IMF = International Monetary Fund, LFS = labor force survey, PSA = Philippine Statistics Authority, STEP = Skills Measurement Program of World Bank.

Sources: Industry employment – PSA, LFS 2017 and ILO; GDP/Output – PSA, McKinsey and IMF Article IV; STEP survey data; Employer survey on impact of 4IR on the electronics industry in the Philippines, n= 41+; Job portal data: jobs in the Electronics industry scraped from the job portal Bestjobs.ph from July to August 2019.

Figure 21: Impact of 4IR on the Level of Skills Required in Electronics Manufacturing, 2018–2030

Electronics: Skills

All skills, with the exception of social, are likely to require significant increases in number of workers with intermediate or advanced level

Absolute Change in Percentage of Workers Requiring Skill at Level, 2018–2030 (%)

Skills	Basic	Intermediate	Advanced
Critical thinking and active learning	(93.6)	72.8	20.8
Written and verbal communication	0.0	(24.5)	24.5
Numeracy	0.0	(100.0)	100.0
Complex problem solving	(82.5)	53.8	28.7
Management	0.0	(68.5)	68.5
Social	(6.4)	(1.7)	8.1
Evaluation, judgment, and decision-making	(0.5)	(99.2)	99.8
Technical	(95.2)	14.0	85.0
Computer literacy	(75.3)	(6.9)	82.2
Digital/ICT skills	(23.0)	89.9	7.4

■ >50% ■ >10% □ ≤10%; ≥ -10% ■ < -10% ■ < -50%

( ) = negative, 4IR = Industry 4.0 or Fourth Industrial Revolution, GDP = gross domestic product, ICT = information and communication technology, IMF = International Monetary Fund, LFS = labor force survey, PSA = Philippine Statistics Authority, STEP = Skills Measurement Program of World Bank.

Sources: Industry employment – PSA, LFS 2017 and ILO; GDP/Output – PSA, McKinsey and IMF Article IV; STEP survey data; Employer survey on impact of 4IR on the electronics industry in the Philippines, n= 41+; Job portal data: jobs in the Electronics industry scraped from the job portal Bestjobs.ph from July to August 2019.

Skills Supply Trends

Figure 22 shows the breakdown of the additional demand for training that will be required by workers in the electronics manufacturing industry under 4IR technology adoption. This reflects the volume of training required to bring the electronics manufacturing workforce in the Philippines from the skills required in 2018 to the level of skills required by 2030, driven only by 4IR technology adoption. Overall, there will need to be 7.5 million additional person trainings by 2030. Similar to IT-BPO, the majority of the required skills development will need to come from on-the-job training. This means that as with the IT-BPO industry, there may be scope for the TVET sector to take a role in delivering shorter, real-world focused training modules within the workplace environment, be it through online and remote learning techniques or by providing in-house trainers to companies.

Figure 22: Additional Person Trainings Required to Meet Skills Demand Driven by 4IR Adoption in Electronics Manufacturing by Training Channel in 2030

Electronics: Skills

The majority of demand for training driven by 4IR adoption will likely need to be serviced by "On-the-job" training

Millions of person trainings required by channel

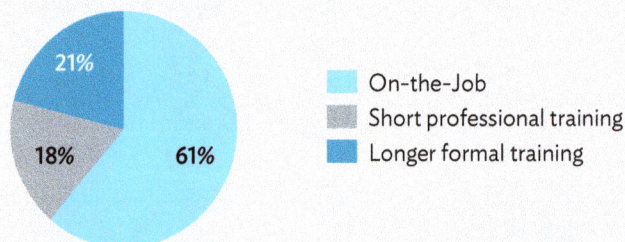

- 21%
- 18%
- 61%

On-the-Job
Short professional training
Longer formal training

4IR = Industry 4.0 or Fourth Industrial Revolution, GDP = gross domestic product, ICT = information and communication technology, IMF = International Monetary Fund, LFS = labor force survey, PSA = Philippine Statistics Authority, STEP = Skills Measurement Program of World Bank.

Notes:
1. Figures include rounding adjustments.
2. One person training refers to training one worker, in one skill from the level required by his occupation's skill profile in 2018 to the relevant level given by the skills profile in 2030.
3. "On-the-job" training refers to training conducted during day to day such as senior staff instructing junior staff or running internal seminars; "Short professional" training refers to short (1 day to 6 months) courses conducted by professional internal or external instructors (e.g., weekend seminars, boot-camps); and "Longer formal" trainings refer to trainings longer than 6 months for which workers would likely have to take leave from their jobs, these include returning into formal education such as obtaining a degree.

Sources: Industry employment – PSA, LFS 2017 and ILO; GDP/Output – PSA, McKinsey and IMF Article IV; STEP survey data; Employer survey on impact of 4IR on the electronics industry in the Philippines, n= 41+; Job portal data: jobs in the Electronics industry scraped from the job portal Bestjobs.ph from July to August 2019.

Employers' opinions about the availability of suitable graduates to fill entry-level roles seem mixed (Figure 23): 40% of surveyed employers agreed that there is a sufficiently large volume of graduates to meet their company's entry-level hiring needs, but 33% disagreed or disagreed strongly. It does appear that this is a problem of volume rather than quality of graduates, as 58% of employers also agreed that graduates were adequately prepared for entry-level jobs and had the right level of general and job-specific skills. This could be especially problematic if one considers the positive net employment impact of 4IR as highlighted earlier in this chapter. If supply of workers is already insufficient, the additional jobs created may add further pressure and workers will need to be sourced from other industries. This will add a further need for retraining, as well as opportunities for TVET.

Figure 23: Employer Sentiment Toward Graduates Hired in the Past 24 Months in Electronics Manufacturing

Electronics: Training

On average employers agree that the quantity and quality of graduates for the electronics industry is sufficient, yet few strongly agree

Respondents (%)

Legend: Don't know | Strongly disagree | Disagree | Neither agree nor disagree | Agree | Strongly agree

Statement	Strongly disagree	Disagree	Neither agree nor disagree	Agree	Strongly agree
There is a large enough volume of graduates from relevant education/training programs to meet my company's entry-level hiring needs	3	30	28	40	0
It is easy to identify and recruit high quality graduates for entry-level positions at my company	0	23	33	45	0
Graduates we hired in the past year have been adequately prepared by their pre-hire education and/or training	0	18	25	55	3
Graduates we have hired in the past year have the appropriate "general" skills including soft and generic skills such as teamwork and creativity		13	30	55	3
Graduates we have hired in the past year have the appropriate "job-specific" skills (e.g., engineering, computer programming)		10	28	58	5
There is a large variance in the quality of vocational secondary graduates depending on region and education provider	5	3	35	58	0

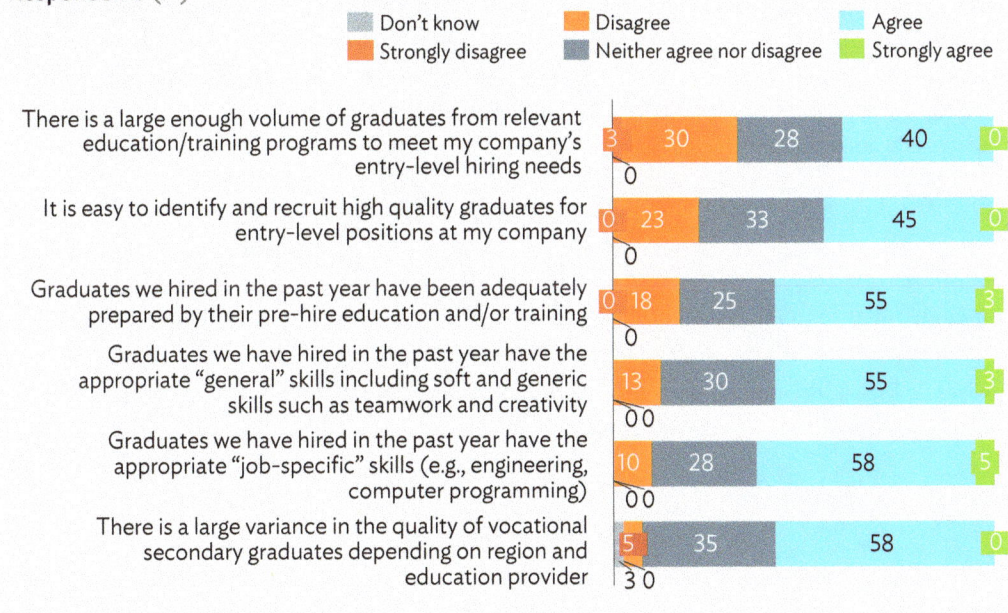

4IR = Industry 4.0 or Fourth Industrial Revolution.

Source: Employer survey on impact of 4IR on the electronics industry in the Philippines, n=40.

CHAPTER 2

Overview of the Training Landscape

This chapter provides insights into the performance of the technical and vocational education and training (TVET) sector in the Philippines as it is preparing to deal with the challenges emerging from Industry 4.0 (4IR) technology adoption. The insights are drawn from a survey of training institutions in the Philippines, complemented with insights from the employer surveys discussed in Chapter 1.

Encouragingly, there is strong alignment between the skills that training institutions believe will be particularly important due to 4IR and the perceptions of employers in the information technology and business process outsourcing (IT-BPO) and electronics manufacturing industries. However, some training institutions may be struggling to keep pace with the rate of change in skills demand. For example, 46% of training institutions surveyed review and update their curricula less than annually, and less than half of institutions provide information on job-market conditions to their students. There also seems to be a severe misalignment between training institutions' expectation of graduate preparedness for work, and employers' expectations about the skills graduates require to perform well in entry-level roles, as well as their general and job-specific skills.

To better understand the supply of talent and skills for the adoption of 4IR, a survey of 120 training institutions was commissioned in the Philippines. The survey focused predominantly on TVET institutions of various levels of schooling. The majority of these are private and, on average, 50% of the funding from the surveyed institutions is from public sources. Institutions of different sizes were sampled, with the smallest training fewer than 100 and the largest training more than 100,000 students annually. The bulk of institutions trained 200–1,000 students annually.

Industry 4.0 Readiness

The majority of institutions feels well prepared for 4IR; however, many request additional technical and financial support (Figure 24). For example, over 80% of institutions believe they have a good understanding of the skills that will be needed to be developed to prepare graduates for their professional lives alongside 4IR technologies. Furthermore, 63% of institutions claim to already have dedicated programs related to 4IR skills, with 83% planning to develop or expand such programs by 2025. While 76% of institutions agree or strongly agree that they will be able to adequately prepare their graduates with their ongoing plans, 88% also agreed or strongly agreed with the statement that they needed additional technical and financial support for dealing with 4IR skill training.

There is strong alignment between the skills that training institutions believe will be particularly important due to 4IR and the perceptions of employers in the IT-BPO and electronics manufacturing industries (Figure 25). The skill category most training institutions deem to become much more important over the next 5 years due to 4IR is technical. This is closely followed by digital and information

Figure 24: Perception of Training Institutions on 4IR Readiness in Electronics Manufacturing

The majority of training institutions generally feel well-equipped for 4IR, however almost 90% request some additional support

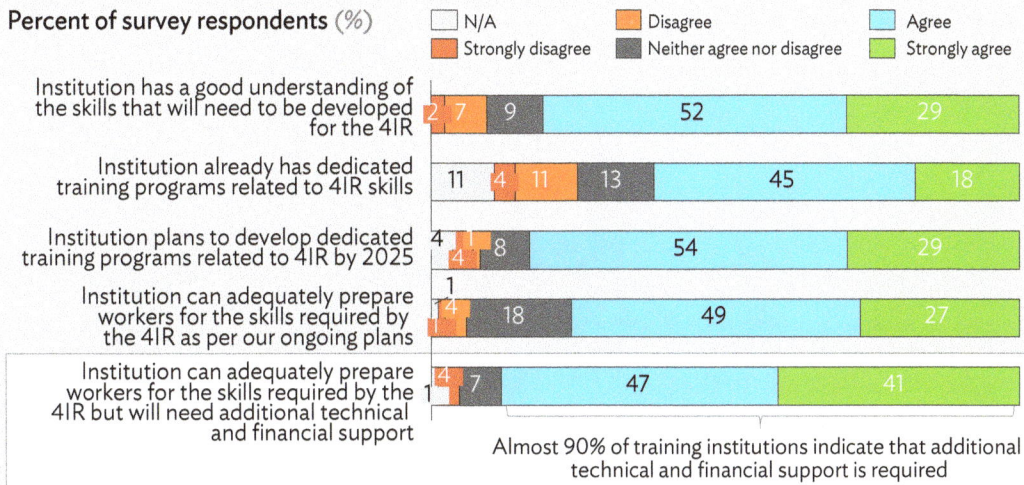

Percent of survey respondents (%)   N/A   Disagree   Agree   Strongly disagree   Neither agree nor disagree   Strongly agree

Institution has a good understanding of the skills that will need to be developed for the 4IR: 2 | 7 | 9 | 52 | 29

Institution already has dedicated training programs related to 4IR skills: 11 | 4 | 11 | 13 | 45 | 18

Institution plans to develop dedicated training programs related to 4IR by 2025: 4 | 1 | 4 | 8 | 54 | 29

Institution can adequately prepare workers for the skills required by the 4IR as per our ongoing plans: 1 | 4 | 18 | 49 | 27

Institution can adequately prepare workers for the skills required by the 4IR but will need additional technical and financial support: 1 | 4 | 7 | 47 | 41

Almost 90% of training institutions indicate that additional technical and financial support is required

4IR = Industry 4.0 or Fourth Industrial Revolution, N/A = not applicable.

Source: Training institution survey on impact of 4IR in the Philippines; n = 85.

Figure 25: Potential Impact of 4IR on the Importance of Different Skills Over the Next 5 Years

Training institutions and employers are generally aligned on which skills will become more important, however, there are differences between service and manufacturing industries

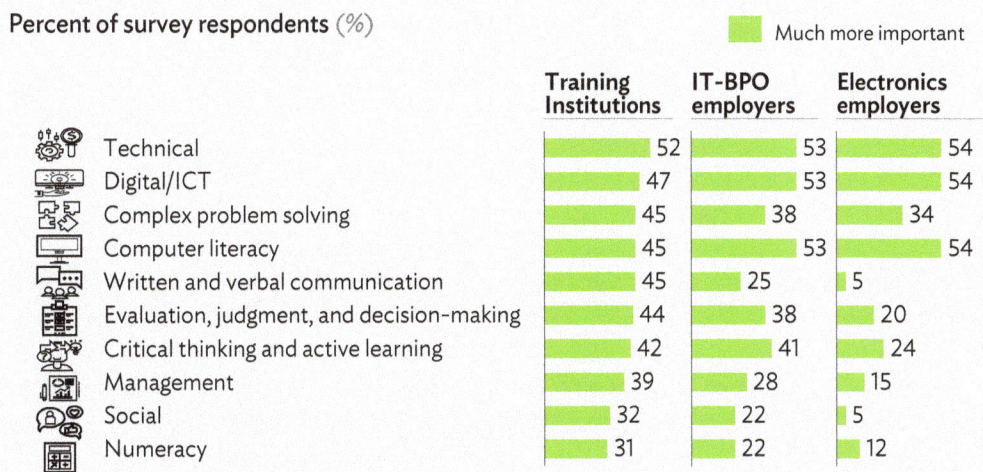

Percent of survey respondents (%)   Much more important

	Training Institutions	IT-BPO employers	Electronics employers
Technical	52	53	54
Digital/ICT	47	53	54
Complex problem solving	45	38	34
Computer literacy	45	53	54
Written and verbal communication	45	25	5
Evaluation, judgment, and decision-making	44	38	20
Critical thinking and active learning	42	41	24
Management	39	28	15
Social	32	22	5
Numeracy	31	22	12

4IR = Industry 4.0 or Fourth Industrial Revolution, ICT = information and communication technology, IT-BPO = information technology and business process outsourcing.

Source: Training institution survey on impact of 4IR in the Philippines, n = 85; Employer survey on impact of 4IR on the IT-BPO industry in the Philippines, n = 32; Employer survey on impact of 4IR on the electronics industry in the Philippines, n = 41.

Figure 26: Frequency of Review and Update of Curricula by Training Institutions

Training Sector: Curriculum

Almost half of all training institutions review and update their curricula less than annually

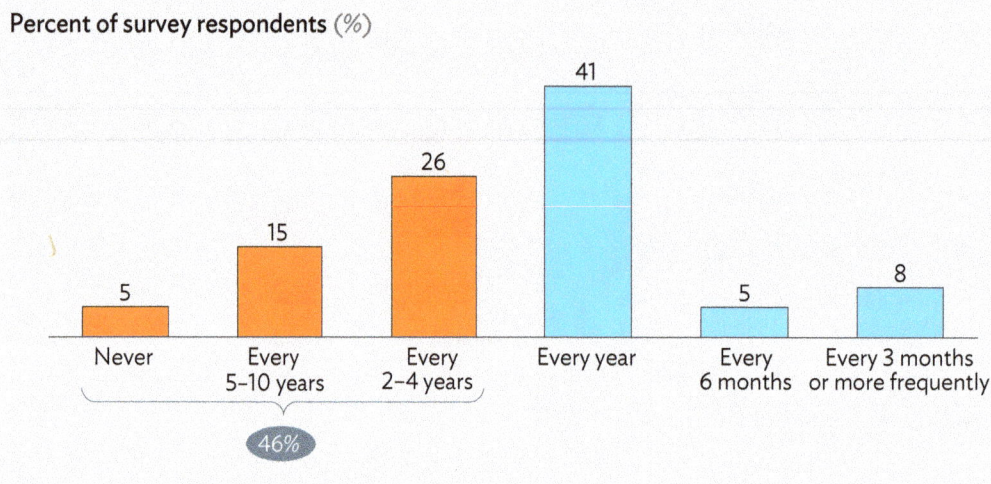

Percent of survey respondents (%)

	Never	Every 5–10 years	Every 2–4 years	Every year	Every 6 months	Every 3 months or more frequently
	5	15	26	41	5	8

46% (Never + Every 5–10 years + Every 2–4 years)

4IR = Industry 4.0 or Fourth Industrial Revolution.

Source: Training institution survey on impact of 4IR in the Philippines; n = 61.

and communication technology (ICT), complex problem solving and computer literacy skills. This ranking is similar to that provided by employers. It is interesting to note that there are some significant differences in skill importance between IT-BPO (which belongs to the service sector) and electronics manufacturing (which belongs to the manufacturing sector). For example, the electronics manufacturing industry does not believe communication and social skills will become much more important under 4IR.

Curricula

Aligning curricula with actual industry needs is one of the most important but often the most challenging components of an effective training and education sector. It relies on frequent updating and close communication with industry, given the speed of change in 4IR technologies in the workplace. Regular curriculum reviews are therefore critical for keeping pace with the skill changes related to 4IR. However, 46% of training institutions surveyed review and update their curricula less than annually (Figure 26).

Another aspect to consider is the content of the curriculum. Countries such as Denmark, Finland, France, Germany, Norway, and Switzerland spend 50%–75% of instructional time at the upper secondary level on practical or on-site training.[18] The results from the Philippines are encouraging in this regard: 43% of time during training is spent in the workplace and an additional 29% on classroom-based student projects. Less than a third of time is allocated to theoretical work (Figure 27).

---

[18]   Organisation for Economic Co-operation and Development. 2010. *Learning for Jobs—The OECD International Survey of VET Systems: First Results and Technical Report.* https://www.oecd.org/education/skills-beyond-school/47334855.pdf

Figure 27: Share of Curriculum Time Spent by Type of Training

Training Sector: Curriculum

There is a lower focus on workplace or practical training than is seen in leading international vocational programs

Average percentage share of total time spent on training type at surveyed institutions (%)

			100
Theoretical	Project-based	Workplace-based	Total time
28	29	43	

According to OECD research, more than three-quarters of vocational training programmes in Denmark, Germany, Finland, France, Norway, and Switzerland at the upper secondary level spend 50%–75% of instructional time in practical or on-site training.

4IR = Industry 4.0 or Fourth Industrial Revolution, OECD = Organisation for Economic Co-operation and Development, VET = vocational education technology.

Note: "Theoretical" training refers to lectures, "Project-based" refers to student projects, and "Workplace-based" refers to on-the-job training such as industry apprenticeships.

Source: Training institution survey on impact of 4IR in the Philippines; n = 61; Małgorzata Kuczera, (2010), Learning for jobs - The OECD International Survey of VET Systems: First Results and Technical Report.

The curriculum in approximately half of the Philippines' training institutions surveyed appear to provide courses relevant for 4IR. When it comes to technology adoption in the classroom, some technologies are heavily favored. However, the actual adoption of 4IR technologies in the classroom by institutions is mixed (Figure 28): 56% of training institutions surveyed stated that they have rolled out courses specifically focusing on 4IR,[19] 56% are also running digital programs to improve digital literacy, and 48% have incorporated additional modules on 4IR-relevant skills into their conventional courses. Of those surveyed, 62% of institutions make use of online self-learning tools and 56% use interactive videos. However, only 33% employ simulators for technical training, 26% use virtual or augmented reality, and only 15% of training institutions have adopted virtual learning platforms. Since the latter technologies are more nascent and expensive, the limited uptake is likely linked to a lack of information about latest applications and/or financial constraints.

---

[19]   Note that in contrast to the questions asked for Figure 11, this portion of the survey focused specifically on the hands-on use of technology in the classroom. Survey respondents were asked if their training institution already offers new courses specifically for 4IR technologies (e.g., education on the use of 4IR technologies in specific industry sectors). Some institutions may have dedicated training programs related to 4IR skills, which could be soft skills (Figure 24), but they have not gone as far as to focus on the use of 4IR technology/machines in courses, e.g., training students to work with or operate robotic manufacturing assistants, etc., or having such technology available for students. Hence, this more nuanced question was included in the survey.

Figure 28: Prevalence of Technology-Related Courses and Technology-Based Delivery in Teaching at Training Institutions

Training Sector: Curriculum

Training institutions provide courses to teach 4IR relevant skills and technologies, but the uptake of 4IR in the classroom is largely limited

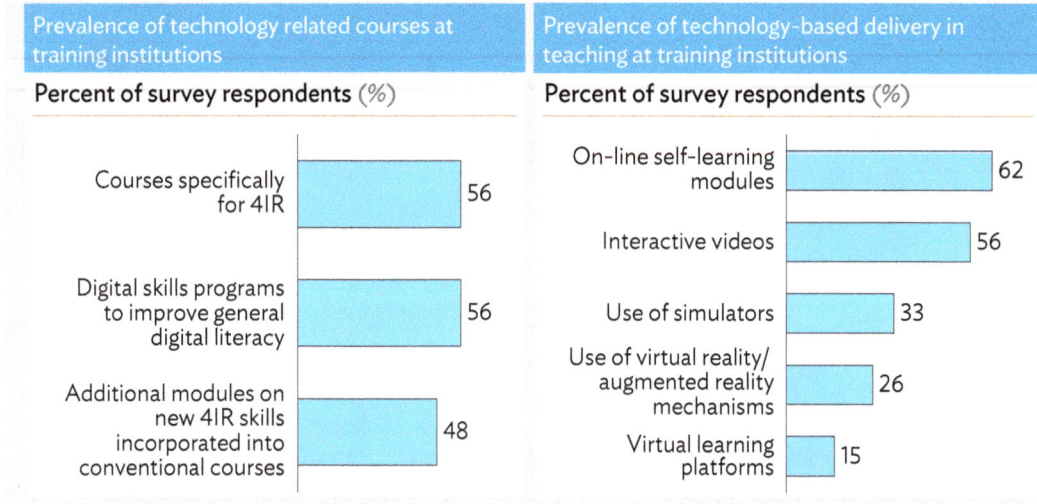

Prevalence of technology related courses at training institutions	Prevalence of technology-based delivery in teaching at training institutions
Percent of survey respondents (%)	Percent of survey respondents (%)

Courses specifically for 4IR	56
Digital skills programs to improve general digital literacy	56
Additional modules on new 4IR skills incorporated into conventional courses	48

On-line self-learning modules	62
Interactive videos	56
Use of simulators	33
Use of virtual reality/ augmented reality mechanisms	26
Virtual learning platforms	15

4IR = Industry 4.0 or Fourth Industrial Revolution.

Source: Training institution survey on impact of 4IR in the Philippines; n = 61.

Many training institutions also engage in a range of programs and activities in addition to training courses, which are aimed at providing students with better information and access to career opportunities and support (Figure 29). For example, on the career front, 80% of institutions surveyed help their students with job applications and interview preparation and 74% arrange visits help with curriculum vitae preparation. Over 60% also provide information on employment and wages of alumni. There is also a significant amount of support to low-income students, as 67% of institutions provide scholarships.

There is some room for improvement since less than half of institutions provide information on job-market conditions (i.e., wages and job prospects in different industries) or job application and interview support. Only 38% say they provide noncareer advice to students such as counseling on personal or financial matters.

Industry Engagement

Training institutions surveyed displayed very positive levels of interaction with potential employers. This was in contrast to anecdotal evidence from targeted interviews with training institutions and from in-country consultation workshops: 45% of training institutions state that they communicate and coordinate with employers in relevant industries several times a year and 30% state that they engage at least a couple of times a year (Figure 30).

Figure 29: Programs Provided in Addition to Training Courses

Training Sector: Curriculum

A number of training institutions provide programs such as job application support, scholarships, and wage information

Percent of survey respondents (%)

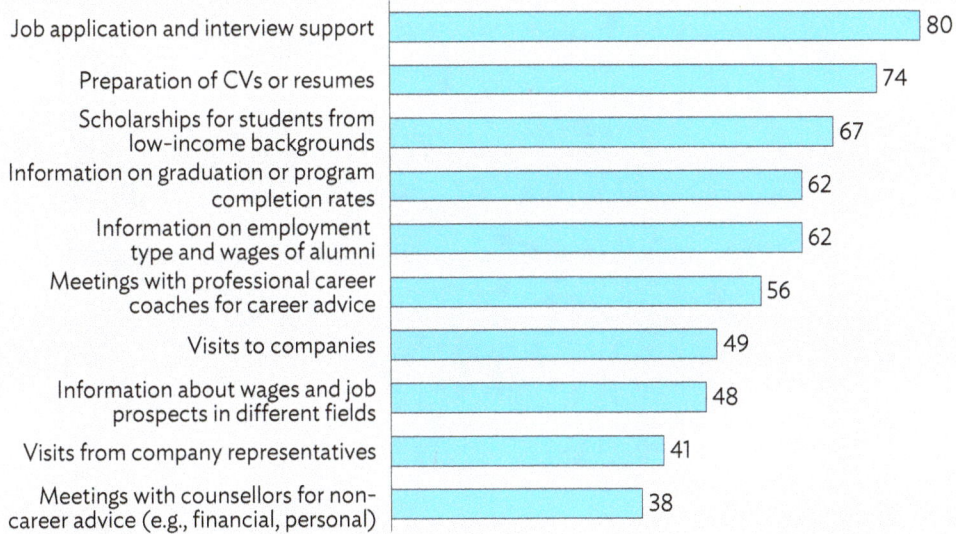

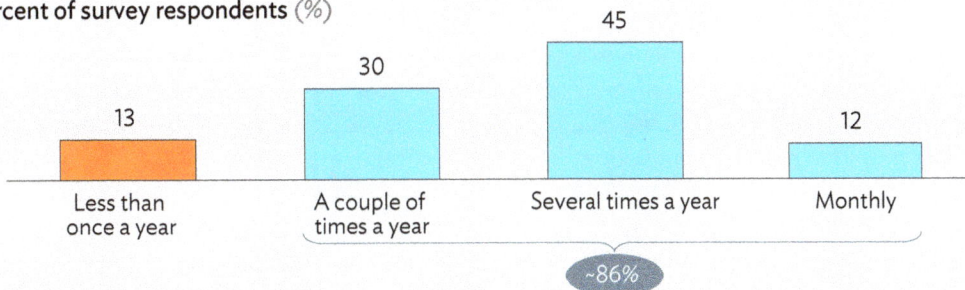

Program	%
Job application and interview support	80
Preparation of CVs or resumes	74
Scholarships for students from low-income backgrounds	67
Information on graduation or program completion rates	62
Information on employment type and wages of alumni	62
Meetings with professional career coaches for career advice	56
Visits to companies	49
Information about wages and job prospects in different fields	48
Visits from company representatives	41
Meetings with counsellors for non-career advice (e.g., financial, personal)	38

4IR = Industry 4.0 or Fourth Industrial Revolution, CV = curriculum vitae.

Source: Training institution survey on impact of 4IR in the Philippines; n = 61.

Figure 30: Frequency of Communication with Employers in Relevant Sectors by Training Institutions

Training Sector: Employers

86% of training institutions communicate with employers at least two times a year

Percent of survey respondents (%)

Less than once a year	A couple of times a year	Several times a year	Monthly
13	30	45	12

~86%

4IR = Industry 4.0 or Fourth Industrial Revolution.

Note: May not add up to 100% due to rounding.

Source: Training institution survey on impact of 4IR in the Philippines; n = 60.

Figure 31: Potential Partnerships and Engagement between Industry and Training Sector

Training Sector: Employers

Training institutions in the Philippines report active engagement with employers with curriculum input and apprenticeships being the most common

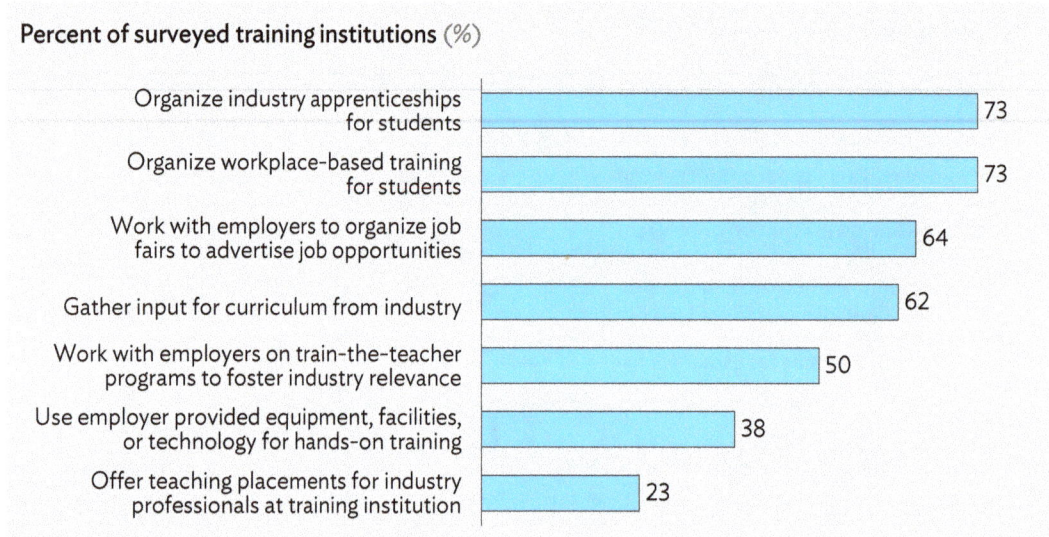

Percent of surveyed training institutions (%)

Organize industry apprenticeships for students	73
Organize workplace-based training for students	73
Work with employers to organize job fairs to advertise job opportunities	64
Gather input for curriculum from industry	62
Work with employers on train-the-teacher programs to foster industry relevance	50
Use employer provided equipment, facilities, or technology for hands-on training	38
Offer teaching placements for industry professionals at training institution	23

4IR = Industry 4.0 or Fourth Industrial Revolution.

Source: Training institution survey on impact of 4IR in the Philippines; n = 56.

Part of the activities included apprenticeships or workplace based training for students organized by 73% of training institutions surveyed. More than 60% indicated that they organize job fairs and gather input for curricula from employers (Figure 31).

While training institutions appear to engage employers actively, active engagement from the side of employers seems to be more limited. For example, only half of the training institutions receive train-the-teacher support from industry, only 38% are able to use employer provided equipment and facilities for hands-on training, and less than 23% offer teaching placements for industry professionals.

While potentially not representative of general employer activity across the industry, these levels of engagement contrast with self-reported activity by employers in the IT-BPO and electronics manufacturing industries (Figure 32). On average, employers in the IT-BPO industry appear more active but employers in both industries appear willing to try deeper engagement and new forms of it.

More than 70% of IT-BPO employers and 80% of electronics manufacturing employers agree or strongly agree with the statement that most of their staff have received training in the past 12 months (Figure 33). The amount of training received can differ by occupation and industry though. For example, administrative workers appear to receive less training regardless of industry. Surprisingly, managerial roles in the electronics manufacturing industry appear to receive significantly less on-the-job and professional training than technical, manual, and customer-facing roles.

Figure 32: Potential Partnerships and Engagement between Industry and Training Sector

Training Sector: Employers

The IT-BPO industry appears more active in engaging the training sector, but employers across the board are willing to explore new options

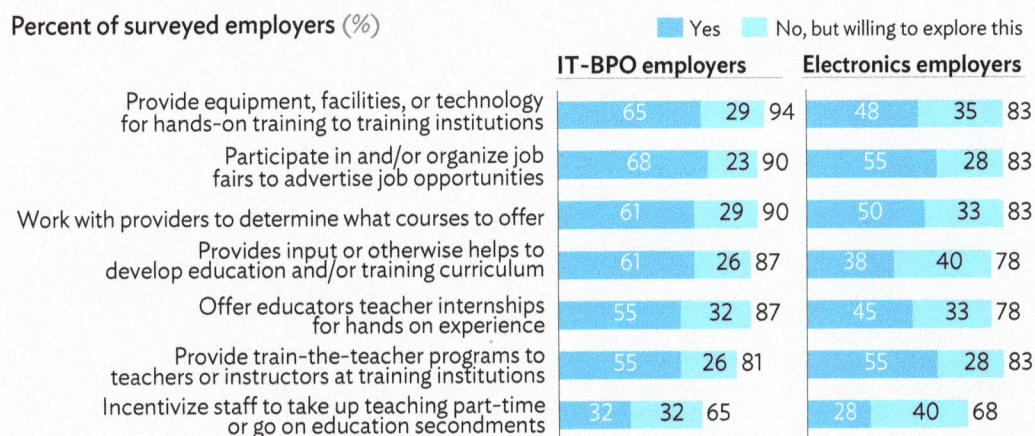

Percent of surveyed employers (%)   ■ Yes ■ No, but willing to explore this

IT-BPO employers | Electronics employers

	IT-BPO Yes	IT-BPO Explore	IT-BPO Total	Electronics Yes	Electronics Explore	Electronics Total
Provide equipment, facilities, or technology for hands-on training to training institutions	65	29	94	48	35	83
Participate in and/or organize job fairs to advertise job opportunities	68	23	90	55	28	83
Work with providers to determine what courses to offer	61	29	90	50	33	83
Provides input or otherwise helps to develop education and/or training curriculum	61	26	87	38	40	78
Offer educators teacher internships for hands on experience	55	32	87	45	33	78
Provide train-the-teacher programs to teachers or instructors at training institutions	55	26	81	55	28	83
Incentivize staff to take up teaching part-time or go on education secondments	32	32	65	28	40	68

4IR = Industry 4.0 or Fourth Industrial Revolution, IT-BPO = information technology and business process outsourcing.

Source: Employer survey on impact of 4IR on the IT-BPO industry in the Philippines, n = 31; Employer survey on impact of 4IR on the electronics industry in the Philippines, n = 40.

Figure 33: Current Annual Training Received while in Employment in IT-BPO and Electronics Manufacturing, by Training Channel

Training Sector: Employers

On average, workers in the electronics manufacturing industry receive fewer training than those in the IT-BPO industry

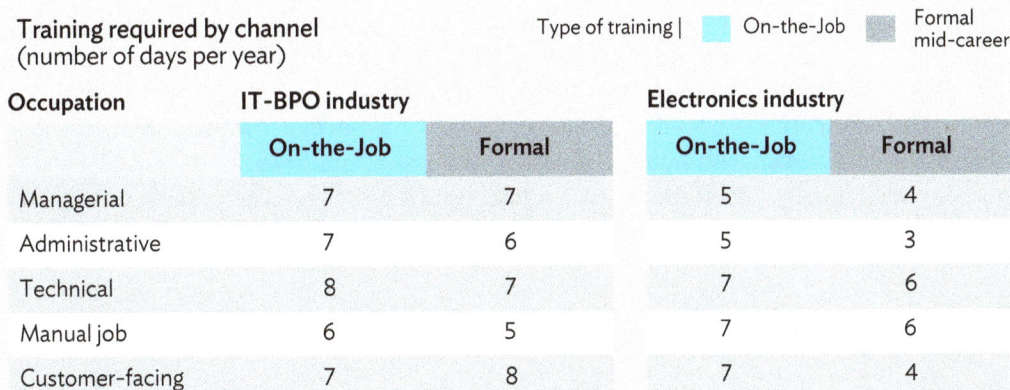

Training required by channel (number of days per year)   Type of training | ■ On-the-Job ■ Formal mid-career

Occupation	IT-BPO industry On-the-Job	IT-BPO industry Formal	Electronics industry On-the-Job	Electronics industry Formal
Managerial	7	7	5	4
Administrative	7	6	5	3
Technical	8	7	7	6
Manual job	6	5	7	6
Customer-facing	7	8	7	4

4IR = Industry 4.0 or Fourth Industrial Revolution, IT-BPO = information technology and business process outsourcing.

Source: Employer survey on impact of 4IR on the IT-BPO industry in the Philippines, n = 31; Employer survey on impact of 4IR on the electronics industry in the Philippines, n = 40.

Figure 34: Practices in Support of Instructors and Teaching Staff of Training Institutions

Training Sector: Instructors

On average, training institutions focus almost equally on both instructor and teacher assessment as well as professional development

Percent of survey respondents (%)

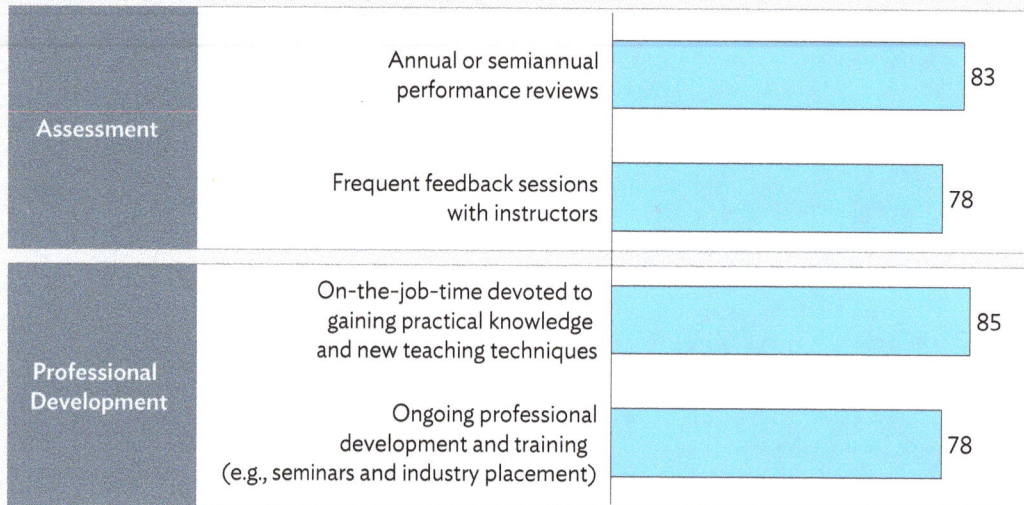

Assessment	Annual or semiannual performance reviews	83
	Frequent feedback sessions with instructors	78
Professional Development	On-the-job-time devoted to gaining practical knowledge and new teaching techniques	85
	Ongoing professional development and training (e.g., seminars and industry placement)	78

4IR = Industry 4.0 or Fourth Industrial Revolution.
Source: Training institution survey on impact of 4IR in the Philippines; n = 60.

Teachers, Trainers, and Instructors

It is encouraging to see that many training institutions are actively engaged in the performance assessment and professional development of their teaching and training staff: more than 80% give formal annual or semiannual performance reviews and 78% provide frequent feedback (Figure 34). Of those surveyed, 85% allow instructors to devote time during working hours to pursue avenues to refresh their practical knowledge and/or learn new techniques on their own and 78% provide ongoing professional development and industry-relevant training (e.g., industry seminars and industry exposure) for instructors. However, only 52% conduct regular visits from company representatives that could help with conveying new techniques to training staff.

Performance and Policy Support

Across their courses, 72% of training institutions face difficulties filling student spots and vacancies (Figure 35). The key reasons appear to be related to lack of price competitiveness, the inability of trainees to differentiate programs, and a lack of knowledge among program trainees.

Figure 35: Reasons for Difficulty in Filling Up Student Spots or Vacancies at Training Institutions

Training Sector: Policy

Over 70% of training institutions find it difficult to fill student vacancies, mostly as a result of a lack of price competitiveness

Percent of survey respondents (%)

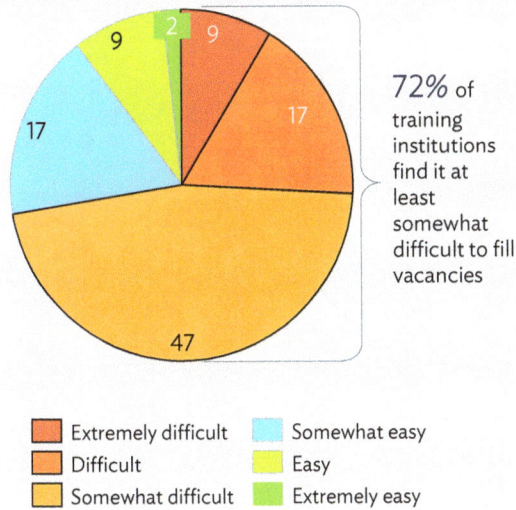

9 · 2 · 9

17

17

47

72% of training institutions find it at least somewhat difficult to fill vacancies

- Extremely difficult
- Difficult
- Somewhat difficult
- Somewhat easy
- Easy
- Extremely easy

Why is it difficult to fill seats at your institution?
Percent of survey respondents with difficulties (%)

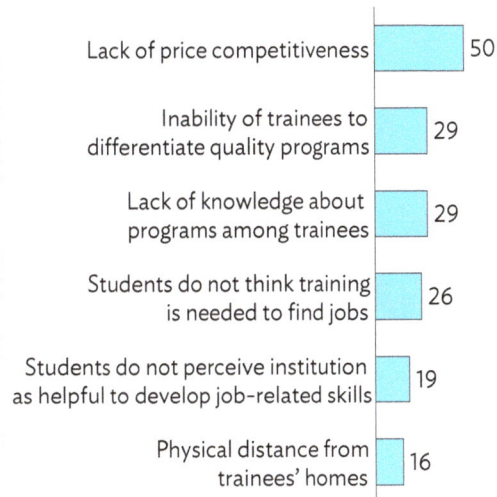

Lack of price competitiveness	50
Inability of trainees to differentiate quality programs	29
Lack of knowledge about programs among trainees	29
Students do not think training is needed to find jobs	26
Students do not perceive institution as helpful to develop job-related skills	19
Physical distance from trainees' homes	16

4IR = Industry 4.0 or Fourth Industrial Revolution.

Source: Training institution survey on impact of 4IR in the Philippines; n = 58.

Training institutions are looking to policy to address some of these challenges (Figure 36). For example, 52% of institutions believe that government inspection and certification processes that assess the quality of educational institutions could have a strongly positive impact. This finding appears consistent with the earlier results that training institutions believe students lack information on the quality of different training providers. To increase training provision, 73% of training institutions cited the importance of government financial support for students (Figure 37).

Supply and Demand Mismatches

According to the training institutions surveyed, the most common reason why graduates may not be able to find a job is that the certifications provided are not well-recognized by employers (Figure 38). The second most common reason is that graduates are not adequately prepared for the jobs they are looking for.

Figure 36: Perception of Policy Effectiveness by Training Institutions

Training Sector: Policy

Policy aimed at quality assessment of institutions is deemed as having the most positive impact by training institutions

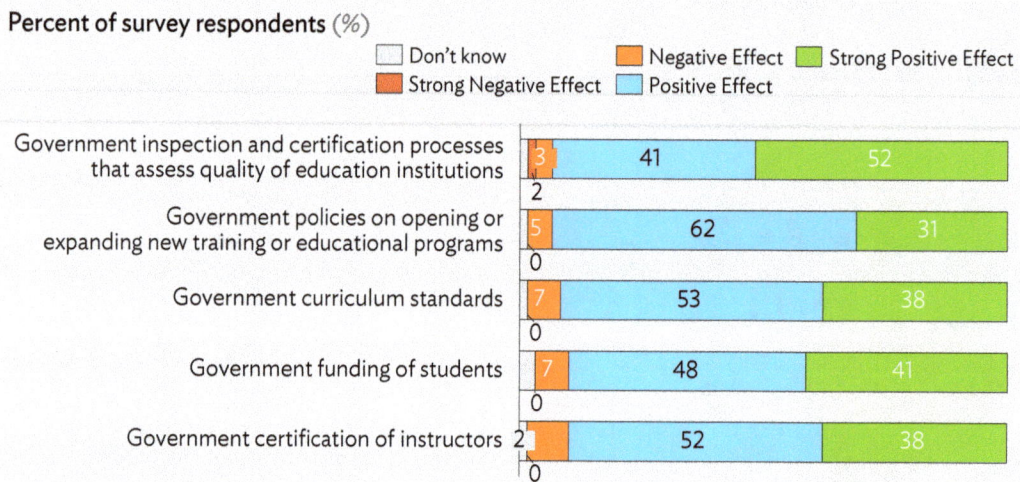

Percent of survey respondents (%)

Legend: Don't know | Negative Effect | Strong Positive Effect | Strong Negative Effect | Positive Effect

Category	Strong Negative	Negative	Positive	Strong Positive
Government inspection and certification processes that assess quality of education institutions	2	3	41	52
Government policies on opening or expanding new training or educational programs	0	5	62	31
Government curriculum standards	0	7	53	38
Government funding of students	0	7	48	41
Government certification of instructors	0	2	52	38

4IR = Industry 4.0 or Fourth Industrial Revolution.

Source: Training institution survey on impact of 4IR in the Philippines; n = 58.

Figure 37: Perspective on Most Impactful Public Policies on Training Provision

Training Sector: Policy

Training institutions believe that government financial support and flexibility on course fees would be the most helpful policies

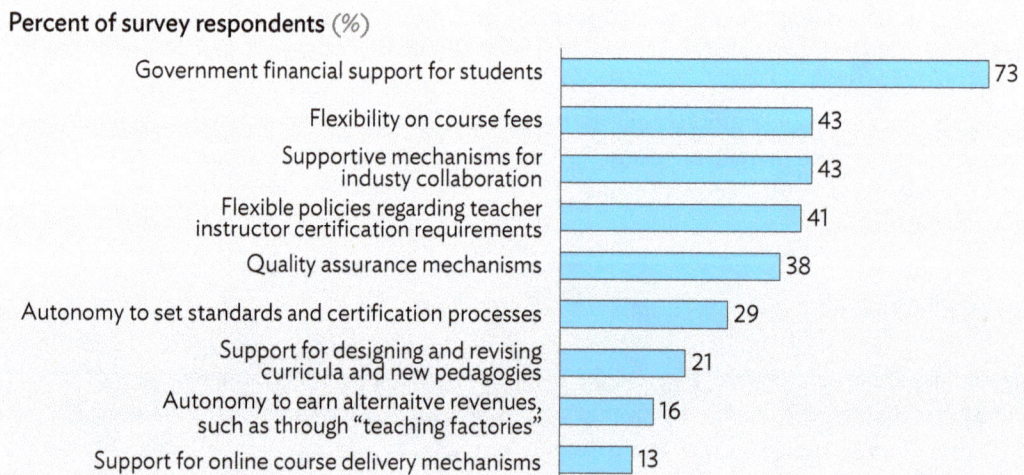

Percent of survey respondents (%)

Policy	%
Government financial support for students	73
Flexibility on course fees	43
Supportive mechanisms for industy collaboration	43
Flexible policies regarding teacher instructor certification requirements	41
Quality assurance mechanisms	38
Autonomy to set standards and certification processes	29
Support for designing and revising curricula and new pedagogies	21
Autonomy to earn alternaitve revenues, such as through "teaching factories"	16
Support for online course delivery mechanisms	13

Source: Training institution survey on impact of 4IR in the Philippines; n = 560.

Figure 38: Reasons for Students Unable to Find Employment Upon Graduation by Prevalence

Training Sector: Students

Training institutions believe a lack of certification recognition and of preparation for jobs by training programs are key employment barriers

Ranking score, 1-Most common; 5-Least common

Rank		Average Ranking
1	Graduates' certifications are not well-recognized by employers	2.5
2	Education and training programs do not adequately prepare job seekers for job opportunities	2.9
3	Not enough job opportunities	3.0
4	Not enough opportunities for job seekers to complete relevant education or training for job opportunities	3.3
5	Enough jobs, but students unaware of job opportunities	3.3

4IR = Industry 4.0 or Fourth Industrial Revolution.

Source: Training institution survey on impact of 4IR in the Philippines; n = 61.

Another area of mismatch between industry and training institutions is around the actual skills, or level of skills (i.e., basic, intermediate, advanced, etc.) that graduates possess when they finish training or education. There seems to be a severe misalignment between training institutions' assessment of graduate preparedness for work, and employers' expectations about the skills graduates require to perform well in entry-level roles, along with their general and job-specific skills (Figure 39).

This significant mismatch in skill expectations between employers and training institutions is particularly surprising given the high reported levels of engagement between employers and training institutions noted earlier. These results suggest that while training institutions may have a good understanding of the skill categories of rising importance for 4IR, the actual implementation of skill training, or depth and specific type of skills taught, do not match industry requirements.

Figure 39: Perception of Preparedness of Graduates for Entry-Level Positions

Training Sector: Students

On average, training institutions are much more optimistic about the preparedness of graduates for work than what employers report

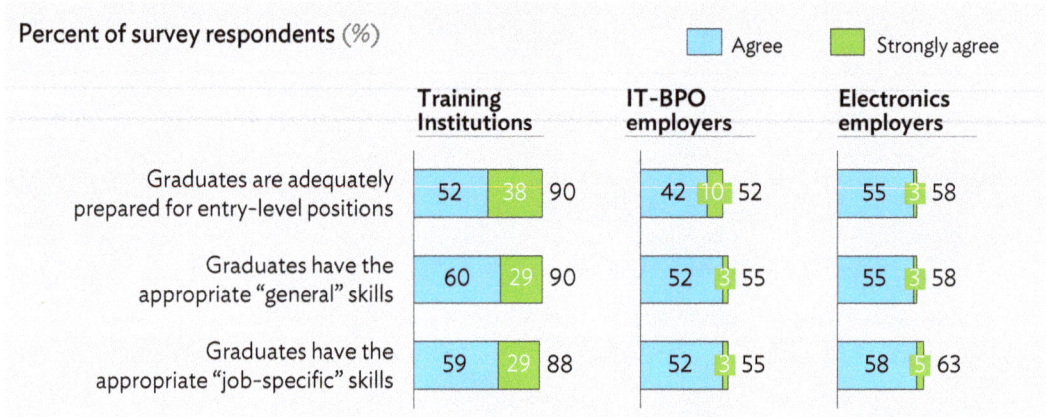

Percent of survey respondents (%)

Agree ■ Strongly agree ■

	Training Institutions	IT-BPO employers	Electronics employers
Graduates are adequately prepared for entry-level positions	52 | 38 = 90	42 | 10 = 52	55 | 3 = 58
Graduates have the appropriate "general" skills	60 | 29 = 90	52 | 3 = 55	55 | 3 = 58
Graduates have the appropriate "job-specific" skills	59 | 29 = 88	52 | 3 = 55	58 | 5 = 63

4IR = Industry 4.0 or Fourth Industrial Revolution, IT-BPO = information technology and business process outsourcing.

Source: Training institution survey on impact of 4IR in the Philippines, n = 58; Employer survey on impact of 4IR on the IT-BPO industry in the Philippines, n = 31; Employer survey on impact of 4IR on the electronics industry in the Philippines, n = 40.

CHAPTER 3

National Policy Responses

A thorough scan of all ongoing policies and programs from the government, industry, and civil society in the Philippines reveals a range of strategies that seek to improve the readiness of the national workforce for Industry 4.0 (4IR). A considerable amount of focus has been placed on fostering collaboration between governments, industry, and civil society to create retraining frameworks for workers. Efforts have also been made to build awareness of jobs and skills that are in demand, establish effective lifelong learning models and develop inclusive programs for underserved groups in the labor market. However, there appears to be a weak focus on a number of critical policy areas. These include creating incentives for employers and workers to participate in skills development, enhancing the relevance of educational curricula to industry needs, and creating social protection mechanisms for flexible workers. While the government has deployed a substantial amount of funding to skills development, this objective could be better achieved by increasing integration between skills policy and 4IR technology adoption strategies, and strengthening incentives for institutions and employers to contribute toward enhancing the responsiveness of human capital to 4IR.

The policy assessment leverages a combination of government policy documents relevant to 4IR and skills, academic literature on the Philippines' skills development, reviews of government policies, and relevant local surveys.

Overview of Industry 4.0 Policy Landscape

4IR has been increasingly viewed as a way to drive new sources of economic growth and international competitiveness.[20] Cognizant of its demographic bonus of a young, well-educated, and relatively information technology (IT)-savvy labor force, the Government of the Philippines views 4IR as an opportunity to drive new growth.[21]

While the government does not currently have one consolidated 4IR strategy, the policies and plans of different ministries reflect increasing awareness of the opportunities and challenges posed by 4IR, as well as initial responses to them. These policies and plans include the following:

(i) **Philippine Development Plan 2017–2022** by the National Economic Development Authority.[22] One of the priorities for the country's economic development outlined in this document is the "acceleration of human capital development." In particular, the number of researchers, scientists, and engineers is set to increase from 270 per 1 million Filipinos to 300 in 2022.

---

[20] Consultations with government agencies of the Philippines in July 2019; J. Kim et al. 2019. Philippine Readiness for the 4th Industrial Revolution: A Case Study. *Asia-Pacific Social Science Review*. http://apssr.com/wp-content/uploads/2019/03/RA-9-R.pdf.

[21] J. Kim et al. 2019. Philippine Readiness for the 4th Industrial Revolution: A Case Study. *Asia-Pacific Social Science Review*. http://apssr.com/wp-content/uploads/2019/03/RA-9-R.pdf.

[22] Government of the Philippines, National Economic and Development Authority. 2017. *Philippine Development Plan 2017–2022*. https://manila2018.dof.gov.ph/wp-content/uploads/2018/01/5-Philippine-Development-Plan-2017-2022.pdf.

(ii) **Comprehensive National Industrial Strategy and Inclusive Innovation Industrial Strategy** by the Department of Trade and Industry (DTI). The Comprehensive National Industrial Strategy (CNIS) aims to position Philippine industries within regional production networks and global value chains, with one of its strategies being to foster innovation-centric networks.[23] A strategy document underpinning the CNIS is the Philippine Innovative Inclusive Industrial Strategy (I[3]S), which lays out a vision and road map for the Philippine economy and its constituent industries to transit to 4IR. Central to this is the implementation of Regional Inclusive Innovation Centers, which are industry-research ecosystems that will be developed to catalyze 4IR technology adoption in the electronics, automotive, and agribusiness industries.[24]

(iii) **National Technical Education and Skills Development Plan 2018–2022 (NTESDP)** by the Technical Education and Skills Development Authority (TESDA).[25] One of its aims is to "prepare the Philippine workforce for the challenges posed by the Fourth Industrial Revolution (4IR) as it ushers in new jobs not yet in the marketplace but that will make existing ones obsolete."[26]

(iv) **Science for Change Program** by the Department of Science and Technology (DOST).[27] Though not specifically geared toward 4IR, the Science for Change Program (S4CP) entails investments in science and technology education, training, and services, with one of the specified objectives being to foster the country's industrial competitiveness.

(v) **National Broadband Strategy** by the Department of Information and Communications Technology (DICT). This aims to improve the affordability, availability, and speed of internet access in the Philippines, which is acknowledged to be lagging behind global peers. The National Broadband Strategy is also key to improving information and communication technology (ICT) infrastructure to effect digital transformation in the economy, and in particular, improve the productive efficiency of businesses.[28]

(vi) **ICT Ecosystem Development** by the DICT.[29] This program aims to develop the Philippines' ICT ecosystem of markets, networks, services, and applications, as well as governance and regulatory frameworks.

Table 1 summarizes the Philippines' key policy documents relating to these areas, with the text in the following chapter further expanding each of them.

---

[23] Government of the Philippines, DTI, and Philippine Board of Investments. 2012. *Comprehensive National Industrial Strategy (CNIS).* http://industry.gov.ph/comprehensive-national-industrial-strategy/; and Government of the Philippines, DTI. 2017. *Philippine Inclusive Innovation Industrial Strategy.* http://boi.gov.ph/sdm_downloads/dti-policy-brief-2017-05-philippine-inclusive-innovation-industrial-strategy/.

[24] Government of the Philippines, DTI. 2017. *Philippine Inclusive Innovation Industrial Strategy.* http://boi.gov.ph/sdm_downloads/dti-policy-brief-2017-05-philippine-inclusive-innovation-industrial-strategy/.

[25] Government of the Philippines, TESDA. 2018. *National Technical Education and Skills Development Plan.* http://www.tesda.gov.ph/About/TESDA/47.

[26] Republic of the Philippines, TESDA. 2018. *National Technical Education and Skills Development Plan.* http://www.tesda.gov.ph/About/TESDA/47.

[27] Government of the Philippines, DOST. 2018. *Guidelines for the Accelerated R&D Program for Capacity Building of Research and Development Institutions and Industrial Competitiveness of the Science for Change (S4C) Program.* http://www.dost.gov.ph/phocadownload/Downloads/Resources/SCIENCE_FOR_CHANGE_PROGRAM_S4CP/S4CP_Guidelines_07Feb2018.pdf.

[28] Government of the Philippines, DICT. 2017. *National Broadband Plan.* https://dict.gov.ph/wp-content/uploads/2017/09/2017.08.09-National-Broadband-Plan.pdf.

[29] More information are available at: Government of the Philippines, DICT. Programs and Projects. https://dict.gov.ph/major-programs-and-projects/.

Table 1: Key Policies Relevant to Managing the Impact of Industry 4.0 on Skills in the Philippines

Policy Document	Responsible Entity	Relevance
Inclusive Innovation Industrial Strategy (I³S)	Department of Trade and Industry (DTI)	Lays out a vision and road map for the Philippine economy for 4IR
National Technical Education and Skills Development Plan (NTESDP) 2018–2022	Technical Education and Skills Development Authority (TESDA)	Sets the priorities for skills development including in relation to 4IR
Small Enterprise Technology and Upgrading Program (SETUP)	Department of Science and Technology (DOST)	National strategy developed to encourage micro, small, and medium-sized enterprises (MSMEs) to adopt technological innovations
Dual Training System Act of 1994	TESDA	Offers tax incentives to companies that provide industry apprenticeships for students—provides a good training platform for 4IR skills
JobStart Philippines	Department of Labor and Employment (DOLE)	Provides workplace based training opportunities for at-risk youth— serves as a good platform to educate 4IR skills to underserved communities
Strengthened Technical and Vocational Education Program (STVEP)	Department of Education (DepEd)	Provides senior high school students with the option to participate in technical and vocational education and training (TVET) through the technical-vocational-livelihood track
Philippine TVET Competency Assessment and Certification System	TESDA	Grants national certificates to recognize workers' skills, assessed based on competency tests

4IR = Industry 4.0 or Fourth Industrial Revolution.

Sources: Government of the Philippines, DTI and Philippine Board of Investments . 2012. *Comprehensive National Industrial Strategy (CNIS)*. http://industry.gov.ph/comprehensive-national-industrial-strategy/; Government of the Philippines, DTI. 2017. *Philippine Inclusive Innovation Industrial Strategy*. http://boi.gov.ph/sdm_downloads/dti-policy-brief-2017-05-philippine-inclusive-innovation-industrial-strategy/; Government of the Philippines, TESDA. 2018. *National Technical Education and Skills Development Plan (2018–2022)*. http://www.tesda.gov.ph/About/TESDA/47; Government of the Philippines, DOST. 2020. Small Enterprise Technology Upgrading Program (SETUP). http://region8.dost.gov.ph/programs-services/technology-transfer-and-commercialization/small-enterprise-technology-upgrading-program.html; International Labour Organization. 2020. Philippines—Dual Training System Act of 1994. https://www.ilo.org/dyn/natlex/natlex4.detail?p_lang=&p_isn=38359&p_classification=09; M. P. M. Lorenzo. 2017. JobStart Philippines: A Promising Project with Some Obstacles. 28 August. https://policyblog.uni-graz.at/2017/08/jobstart-philippines-a-promising-project-with-some-obstacles/; R. C. Alferez. 2012. Implementation of Strengthened Technical Vocational Education Program—Competency Based Curriculum, Northern Mindanao, Philippines. *JPAIR Multidisciplinary Research Journal*. 7 (1). https://ejournals.ph/article.php?id=7497; K. S. Budhrani et al. 2017. *Developing a Skilled Workforce Through Technical and Vocational Education and Training in the Philippines*. https://link.springer.com/referenceworkentry/10.1007%2F978-3-319-38909-7_28-1.

Assessment of Current Philippine Policy Approaches Related to Industry 4.0

A diagnostic approach was taken to understand two important aspects of the Philippines' 4IR policy approach: (i) "the what"—the specific policies being adopted by the Philippines and how they compare to a set of best practice approaches seen internationally in preparing workers for 4IR; and (ii) "the how"—the implementation mechanisms supporting 4IR efforts in government.

Assessment of Policy Actions ("The What")

The country's policies and programs have been grouped into nine action areas assessed to be most crucial to managing the impact of 4IR on jobs and skills.[30] Figure 40 shows the current degree of focus by the country for each action area. The current degree of focus on each action area has been rated as

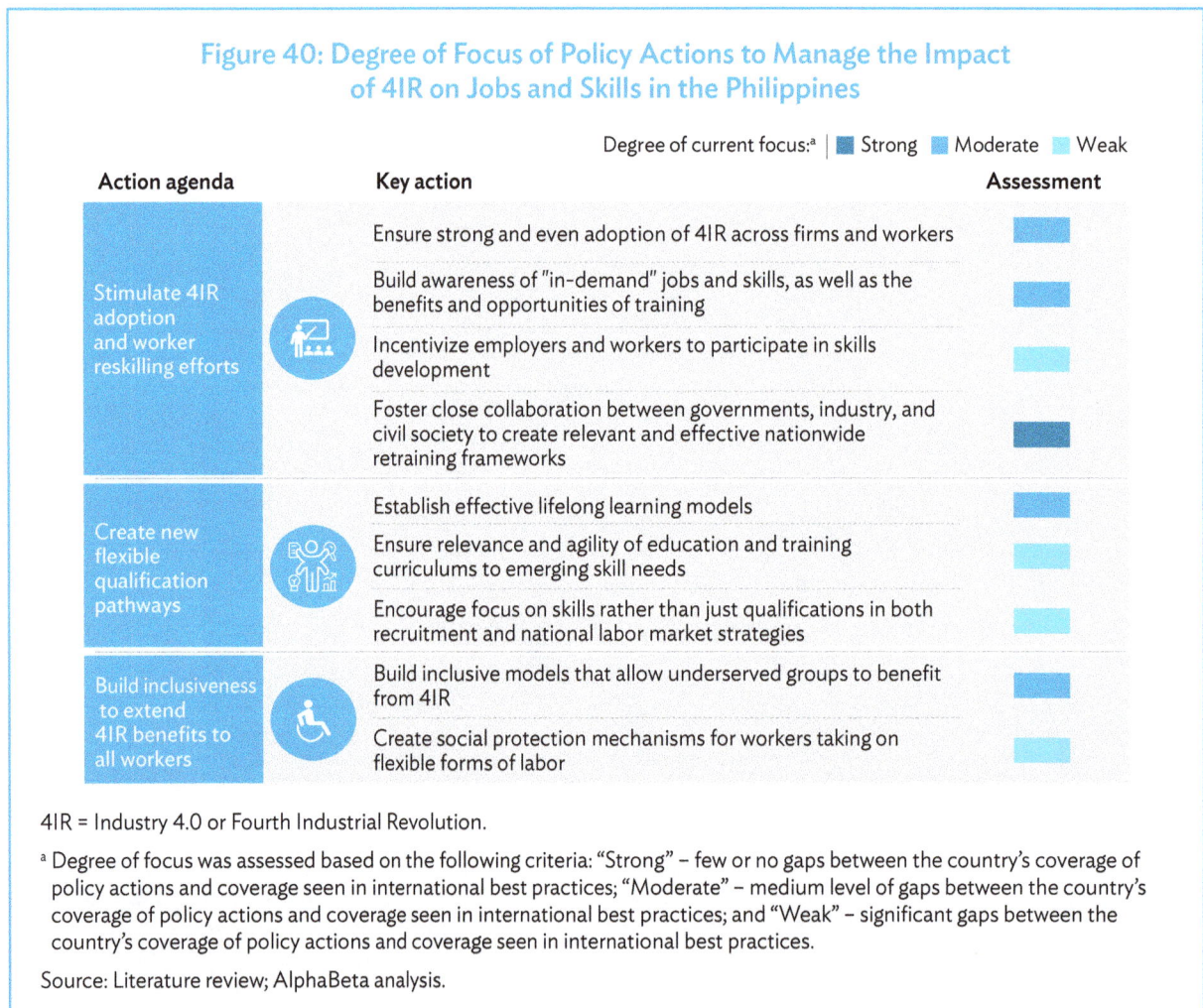

Figure 40: Degree of Focus of Policy Actions to Manage the Impact of 4IR on Jobs and Skills in the Philippines

Degree of current focus:[a]   ■ Strong ■ Moderate ■ Weak

Action agenda	Key action	Assessment
Stimulate 4IR adoption and worker reskilling efforts	Ensure strong and even adoption of 4IR across firms and workers	Moderate
	Build awareness of "in-demand" jobs and skills, as well as the benefits and opportunities of training	Moderate
	Incentivize employers and workers to participate in skills development	Weak
	Foster close collaboration between governments, industry, and civil society to create relevant and effective nationwide retraining frameworks	Strong
Create new flexible qualification pathways	Establish effective lifelong learning models	Moderate
	Ensure relevance and agility of education and training curriculums to emerging skill needs	Weak
	Encourage focus on skills rather than just qualifications in both recruitment and national labor market strategies	Weak
Build inclusiveness to extend 4IR benefits to all workers	Build inclusive models that allow underserved groups to benefit from 4IR	Moderate
	Create social protection mechanisms for workers taking on flexible forms of labor	Weak

4IR = Industry 4.0 or Fourth Industrial Revolution.

[a] Degree of focus was assessed based on the following criteria: "Strong" – few or no gaps between the country's coverage of policy actions and coverage seen in international best practices; "Moderate" – medium level of gaps between the country's coverage of policy actions and coverage seen in international best practices; and "Weak" – significant gaps between the country's coverage of policy actions and coverage seen in international best practices.

Source: Literature review; AlphaBeta analysis.

---

[30]   Based on AlphaBeta research on international best practice for policy actions that manage the impact of 4IR on jobs and skills. Details of these best practices are in Microsoft Asia Stories. 2019. *Preparing for AI: The Implications of Artificial Intelligence for Jobs and Skills in Asian Economies.* https://news.microsoft.com/apac/2019/08/26/preparing-for-ai-the-implications-of-artificial-intelligence-for-jobs-and-skills-in-asian-economies/.

strong, moderate, or weak based on the analyzed extent of the policies' coverage in terms of scope and scale in relation to those observed in international best practice.

Overall, the current degree of focus on the range of 4IR-relevant policy areas is somewhat mixed in the Philippines. While there are strong efforts in fostering multistakeholder collaborations to develop retraining frameworks, there is limited focus on other key areas, particularly around ensuring incentives for employers to invest in the skills development of workers, responsiveness of the curriculum to 4IR-related skills, and developing inclusive models to share the benefits of 4IR. More specifically:

(i) **Stimulating industry adoption and worker reskilling.** A global 4IR readiness assessment indicated that the Philippines was currently in the early planning phase, with the potential to improve its performance in three areas: improving the institutional framework, investing in human capital, and boosting technology platforms.[31] One particular challenge is around raising awareness of in-demand skills. In a recent survey to understand the attitudes of Filipino employers and their workers toward reskilling for artificial intelligence (AI), 45% of employers felt that there were no suitable training programs for their workers, while 48% of workers stated that they did not know what courses to take.[32] Both findings point toward a lack of understanding of the reskilling opportunities in the nation, despite efforts by the government to catalyze them. One example is TESDA's Skills Needs Anticipation survey, carried out to determine employers' desired worker skills and their level of satisfaction with the current competencies of technical and vocational education and training (TVET) graduates.[33] Another is the NTESDP's push to broaden the scope and impact of enterprise-based training, to elevate it as the dominant mode of TVET delivery, particularly through student apprenticeship and dual training systems.[34] However, a challenge in this area relates to incentivizing employers and workers to engage in skills development. A recent survey found that the largest challenge faced by both business leaders and their employees in reskilling for AI was not being able to afford training courses (footnote 32). The employer survey for the electronics manufacturing industry reflected that only 7% of respondents indicated they were able to find the right training providers for their workers' needs, and while the result is more positive for the IT-BPO industry, this number is only one quarter of all employers surveyed. On a more positive note, there have been significant efforts to strengthen linkages between government, academia, and industry, with various policies related to developing training programs for workers and in formulating skills assessment and certification frameworks.

(ii) **Creating new flexible qualification pathways.** Beyond expanding access to higher education and ad hoc training courses, more could be done in the Philippines to equip the country's workforce with the skills necessary for 4IR. Examples include the development of short-term courses in 4IR technologies and the issue of lifetime learning credits. This is particularly crucial as insights from the industry analysis estimate that up to 80% of training needs related to 4IR in the information technology and business process outsourcing (IT-BPO) and electronics manufacturing industries will require on-the-job and short professional training. The Philippines has pursued a range of policies to encourage curriculum responsiveness in both educational and

---

[31] World Economic Forum and A. T. Kearney. 2018. *Readiness for the Future of Production Report 2018.* http://www3.weforum.org/docs/FOP_Readiness_Report_2018.pdf.

[32] Microsoft News Center Philippines. 2018. *Digital Transformation to Contribute US$8 Billion to the Philippines GDP by 2021.* https://news.microsoft.com/en-ph/2018/02/14/digital-transformation-contribute-us8-billion-philippines-gdp-2021/.

[33] Based on consultation with TESDA in January 2020.

[34] Government of the Philippines, TESDA. 2018. *National Technical Education and Skills Development Plan.* http://www.tesda.gov.ph/About/TESDA/47; consultation with TESDA in January 2020.

training institutions to industry needs. An example is TESDA's Adapt and Adopt strategy, under which it benchmarks its TVET programs with internationally recognized standards such as the Dublin, Sydney, and Washington Accords for engineering programs.[35] However, studies reflect significant sources of rigidities in the system, which will have to be alleviated for the smooth functioning of these mechanisms. A study found that, on average, higher education institutions in the Philippines require 18 months to introduce a new degree program, and 13 months to implement curriculum reforms.[36] This is consistent with the insights from the training institution survey in Chapter 2. A deeper analysis reveals four existing sources of rigidity in the Philippines' education system: (i) students often opting for established programs with proven employment results; (ii) the lack of financial resources required to implement curriculum changes and new programs; (iii) the lengthy approval process—assessed to be on average of 1.5 years processing time; and (iv) unrealistic conditions for approval whereby an institution must have already hired the appropriate faculty staff and developed the requisite teaching facilities (footnote 36).

(iii)   **Building inclusiveness to extend the benefits of 4IR to underserved communities.** Strengthening the inclusiveness of skills and education programs is one of the priorities highlighted in the Philippine Development Plan 2017–2022 (footnote 22). Within this is a strong focus on improving the mechanisms for ensuring that students with special needs, from indigenous communities, and who are out of school are provided with the appropriate support to ensure their completion of basic education (footnote 22). However, while there has been significant focus on the inclusiveness agenda, more could still be done to improve the employability of young adults, particularly given that in 2017 almost half of the country's unemployed fell within the age range of 15–24 years, with these youth facing an unemployment rate that is twice the national rate.[37] Another segment of workers to focus on is on-demand or flexible workers—referring to workers who typically find work through online talent platforms or sharing economy applications, and perform tasks for a wide variety of customers.[38] This is significant in the Philippines, where recent studies have found that the country is sixth in the world as the fastest-growing market for the gig economy, is the fifth largest supplier of online labor, and is where at least 2% of the population are freelancers.[39] A recent online survey by PayPal found that 92% of Filipino freelancers cited job stability as a key concern and that 58% of freelancers in four Association of Southeast Asian Nations markets, including the Philippines, have experienced not being paid.[40] However, given the currently limited social protection even for regular workers, there is still some way to achieving this for on-demand or flexible workers in the new economy.

---

[35]   Based on consultation with TESDA in January 2020.

[36]   A. C. Orbeta Jr. et al. 2016. *Are Higher Education Institutions Responsive to Changes in the Labor Market?* https://dirp4.pids.gov.ph/websitecms/CDN/PUBLICATIONS/pidsdps1608.pdf.

[37]   In 2017, the youth unemployment rate was 14.3% compared to the national rate of 5.7%. L. A. Aquino. 2019. Youth Unemployment Still Dominant Problem in Labor Market—Department of Labor and Employment. *Manila Bulletin*. 1 January. https://news.mb.com.ph/2019/01/01/youth-unemployment-still-dominant-problem-in-labor-market-dole/.

[38]   Microsoft. 2018. *The Future Computed: Artificial Intelligence and its Role in Society.* https://blogs.microsoft.com/wp-content/uploads/2018/02/The-Future-Computed_2.8.18.pdf.

[39]   Payoneer. 2019. *The Global Gig-Economy Index: Q2 2019.* https://explore.payoneer.com/q2_global_freelancing_index/; L. Hasnan. 2019. Philippines's Fast-Growing Gig Economy. *The ASEAN Post*. 9 October. https://theaseanpost.com/article/philippines-fast-growing-gig-economy; Oxford Internet Institute. 2019. *The Online Labour Index.* https://ilabour.oii.ox.ac.uk/online-labour-index/.

[40]   Markets include Indonesia, the Philippines, Singapore, and Viet Nam. M. O'Malley. 2018. PayPal Releases Global Freelancer Insights. https://www.paypal.com/stories/us/paypal-releases-global-freelancer-insights; L. Hasnan. 2019. Philippines's Fast-Growing Gig Economy. *The ASEAN Post*. 9 October. https://theaseanpost.com/article/philippines-fast-growing-gig-economy.

Assessment of Implementation of Industry 4.0 Policies ("The How")

Implementation of the Philippines' 4IR strategy for jobs and skills was assessed against three dimensions found to be crucial for success according to past academic work: the clarity and robustness of plans, the strength of coordination between different stakeholders, and the alignment of financing and incentives (Figure 41).[41]

There are a number of implementation challenges across all three dimensions:

(i) **Clarity and robustness of plans.** The starting point for successful implementation is to ensure that there is a vision that is both realistic and clear, and that skills policy is tightly integrated into the overall 4IR strategy. However, a clear 4IR strategy and vision appear to be missing in the Philippines. There is currently no single consolidated 4IR strategy, and national policies very much reside within single government departments. While the DTI's Comprehensive National Industry Strategy and Inclusive Innovation Industrial Strategy establishes the policy for enhancing technology development and research efforts to bring about greater economic productivity, TESDA has a separate NTESDP, which seeks to equip the workforce with the skills to facilitate this transition to 4IR. On the other hand, DOST is leading policies to enhance investments in science and technology education and training. There is much scope for these interrelated policies to be consolidated into a common 4IR road map adopted by the government. Although a memorandum of understanding (MoU) was recently signed by the DTI and the DOST to foster interministerial collaboration on 4IR,[42] beyond this, the government has acknowledged that coordination with other government departments, such as the CHED and individual departments overseeing different industries (e.g., the Department of Agriculture), is limited.[43] There are, nonetheless, intentions to encourage a whole-of-government approach, and a consultation with the DTI revealed a plan by the government to consolidate all 4IR-related policies into a single Philippine Innovation Act.[44] The Implementing Rules and Regulations of this Act was signed in February 2020,[45] creating a Philippine Innovation Council to spearhead all 4IR-relevant efforts.[46]

(ii) **Strength of coordination between different stakeholders.** There is also a lack of coordination across different levels of government. Due to different strategic priorities at both national and local levels of government, national-level targets and plans are at times not prioritized at the grassroots level. A review of the implementation of the JobStart program, for example, found that a key challenge it faced was not being accorded the same level of priority by local government units. This often led to delays in the promised schedules for technical trainings and apprenticeship opportunities under the scheme.[47]

41 Based on AlphaBeta research of 4IR strategies, plus insights from past public sector research, including: M. Barber. 2007. *Instruction to Deliver: Fighting to Transform Britain's Public Services*; McKinsey & Company. 2012. *Delivery 2.0: The New Challenge for Governments.* https://www.mckinsey.com/industries/public-sector/our-insights/delivery-20-the-new-challenge-for-governments.
42 Based on consultation with the DTI in July 2019. Government agencies that were stated to have signed this MoU include DOST and DTI.
43 Government of the Philippines, DTI and Philippine Board of Investments. (Undated). *Industry 4.0: Are We There Yet? I³S Inclusive Innovation Industrial Strategy.* https://pidswebs.pids.gov.ph/CDN/EVENTS/industrialstrategy_aldaba.pdf
44 Based on consultation with the DTI in July 2019.
45 Government of the Philippines, *Philippine News Agency.* 2020. PH Innovation Act IRR Signed. 7 February. https://www.pna.gov.ph/articles/1093284.
46 Based on consultation with the DTI in July 2019.
47 M. P. M. Lorenzo. 2017. JobStart Philippines: A Promising Project with Some Obstacles. 28 August. https://policyblog.uni-graz.at/2017/08/jobstart-philippines-a-promising-project-with-some-obstacles/.

Figure 41: Implementation Challenges Associated with 4IR Policies for Jobs and Skills in the Philippines

Degree of current focus:[a] | ■ Strong ■ Moderate ■ Weak

Dimension	Questions	Assessment
Clarity and robustness of plans	Is there a clearly articulated vision for 4IR?	Weak
	Is there strong integration between employment/skills and the 4IR plan?	Moderate
	Is the plan forward looking, incorporating 4IR trends?	Weak
	Is there strong local data to support evidence-based policymaking?	Weak
Strength of coordination	Is there one shared roadmap across industry and government departments for 4IR?	Weak
	Is there coordination across different government ministries and levels?	Weak
	Is there strong alignment within and between industry, and education and training institutions?	Moderate
Alignment of financing and incentives	Is government financing aligned with the strategic goals?	Moderate
	What are the strength of incentives for employers and workers to invest in skill development? What are the strength of incentives for teachers and institutions to ensure high-quality training and education systems?	Weak

4IR = Industry 4.0 or Fourth Industrial Revolution.

[a] Degree of focus was assessed based on the following criteria: "Strong" – few or no gaps between the country's policy implementation approach and approach seen in international best practices; "Moderate" – medium level of gaps between the country's policy implementation approach and approach seen in international best practices; and "Weak" – significant gaps between the country's policy implementation approach and approach seen in international best practices.

Source: Literature review; AlphaBeta analysis.

(iii) **Alignment of financing and incentives.** For jobs and skills policies to be successful in mitigating the potential negative impacts of 4IR, funding and incentives need to be well-aligned to ensure different stakeholders contribute to skills development. While government financing is aligned with worker skilling priorities, research and development expenditure appears to be low. The incentives for employers and workers to contribute toward skills development also appear weak.

CHAPTER 4

The Way Forward

The previous three chapters highlighted a series of challenges facing the Philippines in relation to Industry 4.0 (4IR). This chapter summarizes those challenges and identifies several recommendations (based on relevant best practice in other countries) for how these could be addressed.

The COVID-19 Effect

The study was undertaken and completed prior to the spread of the coronavirus disease (COVID-19), which has caused unprecedented disruptions to labor markets and to the activities of the workforce across the world. This study's policy recommendations and strategies to strengthen widespread digital capabilities, enhance online or distance learning, digital platforms, education technology, and simulation-based learning have become all the more relevant in the aftermath of COVID-19. The key approaches discussed and elaborated in the report are very relevant in the current context of countries experiencing nationwide closures of schools and training institutions. The expectation is also that after COVID-19, there will be operating procedures that constitute a new normal that entails far more digital capabilities in the workplace. Hence the findings of this study and the follow-on policy directions are crucial and very timely for facilitating a sustainable COVID-19 recovery strategy.

The two sectors chosen for the study in the Philippines, information technology and business process outsourcing (IT–BPO) and electronics, have been extensively and adversely affected by the COVID-19 pandemic. In the IT-BPO sector, there have been widespread disruptions to business operations due to COVID-19. However, the expectation is that there will be lasting shifts in business practices that embody more digital collaborative tools following COVID-19. Similarly, in the electronics industry, the bounce-back after COVID-19 will entail embracing digital supply chains and launching digital sales and marketing initiatives. Hence upskilling and reskilling in 4IR-related occupations is even more urgent for the revival of the economy after COVID-19.

The study obviously does not address the implications of COVID-19 in the Philippines, but the policy directions and future investments for higher-order skills, particularly in the digital domain, are eminently suitable for the country to reimagine new beginnings for the two sectors.

Recap of Industry 4.0-Related Challenges Facing the Philippines

Figure 42 provides a recap of the challenges facing the Philippines from the industry analysis (Chapter 1), the training institution survey (Chapter 2), and the policy assessment (Chapter 3).

Figure 42: Recap of Challenges Facing the Philippines in Relation to 4IR

Area		Key Challenges	Factoids
Industry-level analysis	1	Large displacement of workers in certain industries, with large gender implications	Up to 24% of jobs could be displaced by 4IR technologies in IT-BPO and electronics
	2	Large shift in tasks and skill requirements	Workers in IT-BPO could spend 16% less time on routine physical and routine interpersonal tasks
	3	Significant ramp up of on-the-job training, particularly for analytical skills	Roughly 60% of new trainings related to 4IR will need to be delivered on-the-job
Training institute survey	4	Lack of frequent updates of curriculum	46% of training institutions surveyed review and update their curricula less than annually
	5	Limited adoption of 4IR technologies in the classroom	Only 15% of training institutions are using virtual learning platforms
	6	Mismatch on skill expectations	90% of training institutions believe graduates to be adequately prepared for job market, but only 52% of employers
Policy assessment	7	Lack of flexible skill certification programs	Strong focus on traditional qualifications
	8	Lack of awareness of training opportunities	48% of workers stated that they did not know what courses to take for retraining
	9	Lack of incentives for investment by firms in worker training	Affordability of training courses highlighted as largest barrier for reskilling in the Philippines
	10	Lack of social protection mechanisms for vulnerable workers	<40% of private sector workers contribute to unemployment insurance
	11	Lack of integrated 4IR and skills policy, and coordination between government departments	There is currently no single consolidated 4IR strategy, and national policies very much reside within single government departments

4IR = Industry 4.0 or Fourth Industrial Revolution, IT-BPO = information technology and business process outsourcing.
Source: Asian Development Bank and AlphaBeta.

Recommendations to Address Challenges

There are a number of areas where the Philippines could strengthen its approach to 4IR. Drawing upon international best practice related to the challenges highlighted in Figure 42, several recommendations have been set out to strengthen the Philippines' approach in terms of policy scope and implementation processes (Table 3). For each recommendation, a series of steps or possible approaches have been laid out as a practical road map for implementation. Table 2 shows the key entity suggested as lead for each recommendation, as well as the other stakeholders to be involved. These entities span the government, industry, education, and training sectors, reflecting the importance of strong multistakeholder partnerships for implementing them.

Table 2: Suggested Leads and Stakeholders to Engage for Potential Actions in the Recommendations to Strengthen the 4IR Approach

No.	Recommendation	Key Suggested Lead/s	Stakeholders to Involve
1	Develop 4IR transformation road maps for key sectors	Joint committee constituting the Department of Trade and Industry (DTI), the Department of Science and Technology (DOST), and the Department of Labor and Employment (DOLE)	• Industry associations and relevant sectoral training bodies • Key employers in each industry with experience in 4IR training for their workers • Training institutions • Higher education institutions
2	Develop a series of industry-led technical vocational education and training programs targeting skills for 4IR	Industry associations and key employers in each industry	• DOLE • DTI • Higher education institutions • Training institutions
3	Explore opportunities to increase curriculum responsiveness	Department of Education and Commission on Higher Education (CHED)	• Educational institutions (K-12, TVET, and higher education institutions) • Industry associations and key employers in each industry
4	Upgrade training delivery through 4IR technology in classrooms and training facilities	TESDA, DOST, CHED	• Education technology companies
5	Develop flexible and modular skills certification programs	TESDA and DOLE	• Industry associations (including representatives from companies with strong training programs in each sector) • Training institutions
6	Implement an incentive scheme for firms to train employees for 4IR	DTI	• Industry associations (including representatives from companies with strong training programs in each sector) • Training institutions
7	Formulate new approaches and measures to strengthen inclusion and social protection in the context of 4IR	DOLE	• Employers of freelance workers • Providers of gig economy platforms (e.g., Grab)

4IR = Industry 4.0 or Fourth Industrial Revolution, K-12 = kindergarten and 12 years of basic education, TVET = technical and vocational education and training.

Source: Asian Development Bank and AlphaBeta.

Table 3: Examples of 4IR Skills-Related Best Practices from Around the World

No	Recommendation	Common Challenges	Examples of Countries where Recommendation Implemented
1	Develop 4IR transformation road maps for key sectors	• Lack of understanding of 4IR by businesses • Large displacement of workers in certain sectors, with large gender implications • Limited awareness of training opportunities • Lack of integrated 4IR and skills policy, and coordination between government departments	Australia, Singapore
2	Develop a series of industry-led TVET programs targeting skills for 4IR	• Significant ramp up of on-the-job training, particularly for analytical skills • Mismatch on skills expectations	Denmark, Finland, France, Germany, India, Norway, Switzerland
3	Explore opportunities to increase curriculum responsiveness	• Lack of frequent updates of curriculum	India, Singapore
4	Upgrade training delivery through 4IR technology in classrooms and training facilities	• Limited adoption of 4IR technologies in the classroom	South Africa
5	Develop flexible and modular skill certification programs	• Lack of flexible skill certification programs	Malaysia
6	Implement an incentive scheme for firms to train employees for 4IR	• Lack of incentives for investment by firms in worker training	Malaysia, Singapore
7	Formulate new approaches and measures to strengthen inclusion and social protection in the context of 4IR	• Lack of social protection mechanisms for vulnerable workers	Australia, Japan, Malaysia, Republic of Korea

4IR = Industry 4.0 or Fourth Industrial Revolution, TVET = technical and vocational education and training.

Source: Asian Development Bank and AlphaBeta.

Recommendation 1: Develop Industry 4.0 transformation road maps for key sectors.

Given its implications for economic growth, workforce skills, and technological progress, 4IR strategy and implementation should involve a range of government agencies, as well as industry and training and educational institutions.

While there has been progress in aiming to harmonize approaches—including the recent MoU between the Deparment of Trade and Industry (DTI) and the Department of Science and Technology (DOST) for interministerial collaboration on 4IR,[48] it may be practically challenging to develop a nationwide,

---

[48] Based on consultation with the DTI in July 2019. Government agencies that were stated to have signed this MoU include DOST and DTI.

Box 6: Singapore's Industry Transformation Maps

Singapore's Industry 4.0 (4IR) effort, comprising the industry transformation maps (ITMs), is championed by a dedicated body—the Future Economy Council. Chaired by the Deputy Prime Minister, the Future Economy Council is represented by members from the government, industry, unions, and educational and training institutions.[a] Each ITM represents the road map to 4IR technology adoption for one industry; to ensure coordination and accountability within the government, each ITM is championed by a different government agency, whose area of responsibility is most relevant to the industry.[b] For example, the ITM for the manufacturing sector is led by the Economic Development Agency, while that for the built environment sector is led by the Building and Construction Authority.

The Skills Framework is a key component of the ITMs. Co-created by industry, government, and civil society actors, the framework provides key information on career pathways, the existing and emerging skills required for different occupations, and reskilling options for different sectors. It also provides a list of training programs for skills upgrading. Because of its multistakeholder nature, this framework is also intended to benefit not just workers, but also employers enabling them to identify emerging skills needs for their workers and enhance talent attraction and retention efforts), training providers (allowing them to gain better insights into emerging areas of skill demand and to optimize the targeting of critical skills gaps through appropriate courses), and students (facilitating them to make informed decisions on choice of study based on career aspirations).

A 2018 survey of over 700 firms in Singapore found that 36% of firms take guidance from the ITMs on how to improve their talent pipeline, and how they could address human resources challenges for different sectors.[c]

[a] Government of Singapore, Ministry of Trade and Industry. 2020. The Future Economy Council. https://www.mti.gov.sg/FutureEconomy/TheFutureEconomyCouncil; Government of Singapore, Ministry of Education. 2016. *Formation of the Council for Skills, Innovation and Productivity*. 20 May. https://www.moe.gov.sg/news/press-releases/formation-of-the-council-for-skills--innovation-and-productivity.
[b] Government of Singapore, Ministry of Trade and Industry. 2017. *Media Factsheet—Industry Transformation Maps: Integrated Road maps to Drive Industry Transformation*. https://www.mti.gov.sg/-/media/MTI/ITM/General/Fact-sheet-on-Industry-Transformation-Maps---revised-as-of-31-Mar-17.pdf.
[c] S. K. Tang. 2019. Singapore Businesses Not Investing Enough in Employee Training: SBF Survey: *Channel News Asia*. 17 January. https://www.channelnewsasia.com/news/business/singapore-companies-not-investing-employee-training-sbf-survey-11134230.

Source: Asian Development Bank and AlphaBeta.

cohesive approach. A starting point could be the development of industry transformation maps (ITMs) like in Singapore, which provide information on technology impacts, career pathways, the skills required for different occupations and reskilling options for different industries (Box 6).[49]

The development of these road maps could be led by a joint committee (consisting of DTI, DOST and Department of Labor and Employment [DOLE]), coordinating closely with industry and employer associations as well as training and higher educational institutions. For the IT-BPO industry, this effort could build on the existing work done by the IT and Business Process Association of the Philippines (IBPAP) on studies on the skills landscape for the industry, and communicate recommendations of how to recruit and develop talent for the digital economy to member companies.[50] In the electronics

---

[49] SkillsFuture. 2019. Skills Framework. https://www.skillsfuture.sg/skills-framework.
[50] An example of this study is: IBPAP. 2014. Talent Deep Dive: *An Analysis of Talent Availability for the Information Technology and Business Process Management Industry in 10 Provinces in the Philippines*.

manufacturing industry, this effort could build on the existing engagement between semiconductor and electronics companies with local universities on the skills required by employers for 4IR.[51] Such an approach could also build on the sector-specific insights from the existing Skills Needs Anticipation survey of the Technical Education and Skills Development Authority (TESDA), which (as outlined in Chapter 3) looked into employers' desired worker skills and their level of satisfaction with the current competencies of technical and vocational education and training (TVET) graduates.[52]

Recommendation 2: Develop a series of industry-led technical vocational education and training programs targeting skills for Industry 4.0.

The poor quality of TVET programs and their lack of industry relevance was identified in the employer surveys (Chapter 1) and in the training institution survey (Chapter 2). This is in spite of several ongoing efforts in the IT-BPO and electronics manufacturing industries. The IBPAP routinely carries out studies of the skills landscape for the sector, and communicates recommendations on how to recruit and develop talent for the digital economy to member companies.[53] It also collaborates with training institutions and government agencies to embed curricula and courses within private and public TVET curricula deemed critical by IT-BPO employers for 4IR.[54] This level of engagement has also been seen in the electronics manufacturing industry, where companies have partnered with local universities to shape curricula to be more relevant to industry needs.[55]

To strengthen their quality and relevance, it is recommended that industry associations take the lead in developing a series of TVET programs. They could be supported by DOLE and DTI, and the industry associations could also work together with training and educational institutions to scope such programs. Parameters to be designed include training curricula (especially which 4IR technologies are to be taught) and their durations, teacher recruitment and training, and applicant criteria. In firming these parameters, the formats of industry-led TVET programs in international best practice may be considered. Some notable examples include industry bootcamps by McKinsey & Company's Generation program, which operates across several countries (Box 7).[56]

Recommendation 3: Explore opportunities to increase curriculum responsiveness.

Analysis in Chapter 3 highlighted the fact that while there are mechanisms in the Philippines to incorporate curriculum changes to meet industry needs, present sources of rigidity restrict the speed with which these updates can take place. This is detrimental in a landscape of rapidly evolving technologies and associated skills needs.

Ongoing mechanisms to ensure alignment between curricula and industry needs in both the IT-BPO and electronics manufacturing sectors include: partnership between the IBPAP and the Commission on Higher Education (CHED) to develop a systems thinking course as part of tertiary curricula;[57] and direct partnerships between individual companies and universities (such as that between local electronics manufacturing firm EMS Components Assembly and De La Salle University to shape the curriculum

---

51   Based on consultation with EMS Components Assembly Inc in July 2019.
52   Based on consultation with TESDA in January 2020.
53   An example of this study is: IBPAP. 2014. *Talent Deep Dive: An Analysis of Talent Availability for the Information Technology and Business Process Management Industry in 10 Provinces in the Philippines.* Provided by the IBPAP.
54   Based on consultation with the IBPAP in July 2019.
55   Based on consultation with EMS Components Assembly Inc in July 2019.
56   Asia Philanthropy Circle. 2017. *Catalysing Productive Livelihood: A Guide to Education Interventions with an Accelerated Path to Scale and Impact.* http://www.edumap-indonesia.asiaphilanthropycircle.org/wp-content/uploads/2017/11/APC-Giving-Guide-Book-Final-Report-17112017.pdf.
57   Based on consultations with the IBPAP in July 2019.

Box 7: Connecting Students and Industry through Boot Camps

Industry needs appropriately trained recruits, and youth job seekers need to be hired. Industry boot camps can help connect the skills offered by young job seekers to those needed by industry. The Generation program develops programs focusing on four sectors, with teaching facilities in 119 cities in six continents. The program is offered to 18- to 29-year-olds.[a] Among the program's features are direct contact with potential employers, matching trainee attributes with employer needs, courses that cover technical, behavioral, and mental skills, continuous monitoring and support during and after the program, and a strong alumni network.

Since its inception, 31,600 people have gone through the training, with 80% finding jobs within 3 months of finishing the program and 65% of those staying with their jobs for at least a year (footnote a). Employers also rate program graduates as higher performing than others.[b]

[a] Generation program. Undated. https://www.generation.org/.
[b] Asia Philanthropy Circle. 2017. *Catalysing Productive Livelihood: A Guide to Education Interventions with an Accelerated Path to Scale and Impact.* http://www.edumap-indonesia.asiaphilanthropycircle.org/wp-content/uploads/2017/11/APC-Giving-Guide-Book-Final-Report-17112017.pdf.

Source: Asian Development Bank and AlphaBeta.

for its electronics engineering courses).[58] Despite these, the survey of training institutions, as outlined in Chapter 2, reflected that they may be struggling to keep pace with the rate of change in skills demand; the survey also highlighted an apparent misalignment between the institutions and employers' expectations about the skills required by graduates to perform well in entry-level roles.

It is recommended that the government departments overseeing educational curricula—the Department of Education (DepEd) and CHED—work closely with educational institutions to properly understand the reasons for time lags and find ways to address them in curriculum changes. Based on the bottlenecks identified, there could be a range of potential actions to address them, including joint industry–institution branding exercises for new courses; the establishment of a fund and in-kind support for the setup costs of new courses; and streamlining approval processes and conditions for curriculum changes. DepEd and CHED should also engage closely with private sector employers to understand some of the bottlenecks faced in reflecting industry needs in curricula.

Recommendation 4: Upgrade training delivery through Industry 4.0 technology in classrooms and training facilities.

An effective way of preparing students or future workers for 4IR—or at least equipping them with the basic digital literacy skills necessary to excel in the future economy—is to apply 4IR technologies in classrooms. Artificial intelligence (AI), for example, has been used to stimulate critical thinking through a virtual environment for building and assessing higher-order inquiry skills.[59] AI-enabled immersive computer games, for instance, have been used for science, technology, engineering, and mathematics (STEM) education in some schools in the United States.[60] Personalized learning approaches adopting

58 Based on consultations with EMS Components Assembly in July 2019.
59 J. M. Spector and S. Ma. 2019. Inquiry and Critical Thinking Skills for the Next Generation: From Artificial Intelligence Back to Human Intelligence. *Smart Learning Environments.* 9 (8). https://slejournal.springeropen.com/articles/10.1186/s40561-019-0088-z.
60 D. Jass Ketelhut et al. 2009. A Multi-User Virtual Environment for Building and Assessing Higher Order Inquiry Skills in Science. *British Journal of Educational Technology.* https://onlinelibrary.wiley.com/doi/abs/10.1111/j.1467-8535.2009.01036.x.

Box 8: Leveraging Technologies to Improve Education and Manage Gender Gaps

Digital technologies have been shown to improve education quality and even manage gender gaps from an early age.

The African School for Excellence, an affordable private secondary school in South Africa, deploys an innovative rotational classroom model in which students rotate between teacher-facilitated lessons, small-group peer-learning activities and individual work on computers supervised by trainee teachers with online courses from free products such as Khan Academy.[a] This blended learning approach innovatively reduces costs by using a smaller number of highly trained teachers while enhancing education outcomes with its emphasis on personalized learning and small class sizes. Students in the African School for Excellence have been found to outperform the wealthiest students in the country by 2.3 times in mathematics and 1.4 times in English. At the same time, the per-student cost of US$800 a year is low compared with South African averages, which are in the range of US$1,400 to US$16,500 per year (footnote a).

Such personalized adaptive learning digital tools are also beginning to show their potential in bridging gender differences in students' attainment from a young age. onebillion, a London-based nonprofit organization focused on building scalable educational software for children, launched the app onecourse, which delivers content and practice on a tablet.[b] This app was found to prevent a gender gap in reading and mathematics skills from surfacing among first-grade students in Malawi—potentially by overcoming sociocultural factors responsible for gaps emerging in traditional classroom settings.[c]

[a] Center for Universal Education at Brookings. 2019. *Learning to Leapfrog: Innovative Pedagogies to Transform Education.* https://www.brookings.edu/wp-content/uploads/2019/09/Learning-to-Leapfrog-InnovativePedagogiestoTransformEducation-Web.pdf.
[b] onebillion. Undated. *onecourse: one app that delivers reading, writing and numeracy.* https://onebillion.org/onecourse/app/
[c] Pathways for Prosperity Commission. 2019. *Positive Disruption: Health and Education in the Digital Age.* https://pathwayscommission.bsg.ox.ac.uk/positive-disruption.

Source: Asian Development Bank and AlphaBeta.

digital channels in which students access customized learning content through online education programs could also be incorporated into formal curricula. As shown in the training institution survey results in Chapter 2, however, technology adoption in the classroom in the Philippines is currently limited in many institutions.

This is an area in which DOST, TESDA, DepEd, and even CHED could jointly lead, working with education technology firms in the country to incorporate 4IR technologies in classrooms from elementary schools all the way through to universities and polytechnics. As installing such technologies and equipment at every institution could be financially challenging, a possible way to reduce the overall cost would be to adopt blended learning approaches. These combine classroom and personalized online learning and have been demonstrated to be highly effective at not just improving education outcomes but also managing gender inequality (Box 8).

4IR technologies that are already in use in the IT–BPO and electronics manufacturing industries can also be taught directly in classroom environments. For instance, students could be familiarized with chatbot technologies (increasingly used in the IT–BPO industry) and with the processes through which chatbots become more efficient at answering common queries. Those training for the electronics manufacturing industry could also be taught about computerizing manufacturing process flows through new software.

Box 9: The Malaysian Skills Certification Program

In Malaysia, individuals who do not possess formal educational qualifications can enter their desired careers through the Malaysian Skills Certification Program. Recognized by industry, this program awards skills certificates at five different levels:

(i)   Malaysian Skills Certificate (SKM) Level 1
(ii)   Malaysian Skills Certificate (SKM) Level 2
(iii)   Malaysian Skills Certificate (SKM) Level 3
(iv)   Diploma in Skills Malaysia (DKM) Level 4
(v)   Malaysian Skills Advanced Diploma (DLKM) Level 5[a]

These certificates are awarded across all 22 sectors of the economy according to the country's National Occupational Skills Standard.[b] Importantly, no formal educational qualifications are required—the only requirements for candidates are the ability to speak and write in both Bahasa Melayu and English, and a pass in a lower Skills Certificate level before being able to qualify for a higher level in the same field (footnote a).

Candidates may obtain these certificates through three channels: training in institutions accredited by the Jabatan Pembangunan Kemahiran (Department of Skills Development); industry apprenticeships under the National Dual Training System; and Accreditation of Prior Achievement (footnote a). The third channel refers to accreditation gained through past work and/or training experience.

With these certificates being accredited as officially recognized qualifications and mapped to equivalent academic qualifications under the Malaysian Qualifications Framework, Malaysian companies are able to assess the suitability of candidates without formal education who possess the relevant skills to excel at a job.[c]

[a] Government of Malaysia, Department of Skills Department. Undated. Malaysian Skill Certificate (SKM). https://www.dsd.gov.my/jpkv4/index.php/en/malaysian-skills-certificate.
[b] Organisation for Economic Co-operation and Development. 2012. *Skills Development Pathways in Asia*. https://www.oecd.org/cfe/leed/Skills%20Development%20Pathways%20in%20Asia_FINAL%20VERSION.pdf.
[c] Government of Malaysia, Ministry of Higher Education and Malaysian Qualifications Agency. 2019. Malaysian Qualifications Framework. https://www.mqa.gov.my/pv4/mqf.cfm.

Source: Asian Development Bank and AlphaBeta.

Recommendation 5: Develop flexible and modular skill certification programs.

As outlined in Chapter 3, there is still a strong emphasis on traditional qualifications attained through the education system or competency assessments. Even experience-based accreditation in the Philippines requires the applicant to have at least graduated from high school. It is recommended that the Philippines explore the development of flexible skills certification programs that recognize skills attainment outside traditional education.

A positive example of a skills-based accreditation system is the Malaysian Skills Certification Program, in which skills certificates are given to workers who do not have any formal educational qualifications but who have obtained the relevant knowledge, experience and skills in the workplace to enhance their career prospects (Box 9).[61]

---

[61]   Organisation for Economic Co-operation and Development. 2012. *Skills Development Pathways in Asia*. https://www.oecd.org/cfe/leed/Skills%20Development%20Pathways%20in%20Asia_FINAL%20VERSION.pdf.

In the Philippines, it is recommended that TESDA and DOLE review current certification frameworks to assess if educational qualifications (e.g., the Expanded Tertiary Education Equivalency and Accreditation scheme) and competency assessments (e.g., as required for National Certificates) may be augmented with more flexible skills certification programs. A starting point could be for TESDA and DOLE to work with industry and training institutions to review the effectiveness of minimum educational qualifications and competency-based assessments in current certification frameworks, particularly in the IT-BPO and electronics manufacturing industries. The review should be based on the practical need for such qualifications, and the extent to which such qualifications adequately represent specific skills competency levels.

Recommendation 6: Implement an incentive scheme for firms to train employees for 4IR.

Despite the substantial productivity gains, that 4IR technologies could bring about (as demonstrated in Chapter 1), employer-led training efforts in the Philippines remain limited. A recent survey showed that 4% more employers in the Philippines would rather hire new staff with the required skills than retrain existing workers, and that 74% of employers simply expect employees to pick up skills on the job.[62] This could be explained by three key drivers:

(i)   There are information asymmetries related to a limited awareness of the new skills that are required for 4IR and the relevant courses for them. In a recent survey to understand the attitudes of Filipino employers and their workers toward reskilling for AI, 45% of employers felt that there were no suitable training programs for their workers to take, while 48% of workers stated that they did not know what courses to take (footnote 32).

(ii) There appears to be a lack of well-functioning markets for training services in the Philippines, characterized by a shortage of high-quality training courses and instructors as well as information asymmetries about available training options. The employer survey for the electronics manufacturing and the IT-BPO industries reflected that only 7% and a quarter of respondents, respectively, indicated they were able to find the right training providers for their workers' needs.

(iii) While Philippine firms are aware of the need to develop training plans for their employees, many lack the budgets to implement them. A recent survey found that the largest challenge faced by both business leaders and their employees in reskilling for AI was not being able to afford training courses (footnote 32).

Given all of these, it is critical to develop a set of support programs to encourage firms to invest in relevant 4IR training for their workers. This action would involve developing appropriate incentive programs for firms to invest in worker skill development related to 4IR.

This action could be potentially led by DTI in consultation with industry associations (including representatives from companies with strong training programs in each sector) and training institutions. The four steps would be:

(i)   Identify appropriate incentive programs for firms (Box 10 provides some international examples);

(ii) Undertake a holistic cost–benefit analysis of the incentive schemes and associated training programs (noting that cost–benefit analyses of skills training programs done from the

---

62   World Economic Forum. 2018. *The Future of Jobs Report 2018*. http://www3.weforum.org/docs/WEF_Future_of_Jobs_2018.pdf.

> ### Box 10: Incentive Schemes for Firm Training in the Region
>
> The Government of Singapore provides subsidies for employee training course fees and absentee payroll salary costs, with higher incentives being awarded for courses that are government certified.[a] For example, while subsidies for both government-certified and approved certifiable courses cover 90%–95% of course fees, those for the latter have hourly caps. On the other hand, the subsidies for noncertifiable courses are lower, at US$2 per hour of training. Absentee payroll funding covers up to 95% of the hourly basic salary. The Government of Malaysia has a similar program known as the Skills Upgrading Program, which provides grants covering 70% of training fees for technical and soft skills to small and medium-sized enterprises.[b]
>
> [a] SkillsFuture SG. Funding Support for Employers. https://www.ssg.gov.sg/programmes-and-initiatives/funding/funding-for-employer-based-training.html.
> [b] Microsoft and AlphaBeta. 2019. Preparing for AI: The Implications of Artificial Intelligence for Jobs and Skills In Asian Economies. https://news.microsoft.com/apac/2019/08/26/preparing-for-ai-the-implications-of-artificial-intelligence-for-jobs-and-skills-in-asian-economies.
>
> Source: Asian Development Bank and AlphaBeta.

government's perspective have tended to focus largely on the direct economic costs of the program, while disregarding indirect economic benefits such as reduced welfare payments due to lower unemployment rates resulting from the training);

(iii) Set up pilot programs in a number of priority industries (the IT-BPO and electronics manufacturing industries could be starting points); and

(iv) Scale the programs to other industries, incorporating lessons learned from the pilot.

Recommendation 7: Formulate new approaches and measures to strengthen inclusion and social protection in the context of Industry 4.0.

The lack of social protection policies for the rapidly growing number of on-demand workers in each country was highlighted as a key concern in Chapter 3. It is recommended that cost–benefit analyses of several policy options for the social protection of such workers be conducted and potentially pilot schemes could be set up to test their broader applicability. These could include exploring policy approaches to enhance the income security for on-demand workers (e.g., in Australia, workers on short-term contracts are entitled to an increment of 25% for each hour worked compared to a worker doing the same job on an ongoing basis).[63] Another option could be working with key employers to champion corporate policies mandating income stability for on-demand workers. There are some innovative approaches that could be used for this. For example, Care.com, a platform for caregivers to find work, enables families employing such services to contribute to their caregiver's benefits similar to how traditional corporate employers fund employee benefits.[64] DOLE could lead this initiative and work closely with companies that make use of freelance labor as well as the providers of online freelance work platforms.

---

[63] Organisation for Economic Co-operation and Development. 2018. *The Future of Social Protection: What Works for Non-Standard Workers?* https://doi.org/10.1787/9789264306943-en.

[64] G. Bonoli. 2019. Ensuring economic Security in the Gig Economy. *The Business Times.* 13 March. https://www.businesstimes.com.sg/opinion/ensuring-economic-security-in-the-gig-economy; Microsoft. 2018. *The Future Computed: Artificial Intelligence and its Role in Society.* https://blogs.microsoft.com/wp-content/uploads/2018/02/The-Future-Computed_2.8.18.pdf.

Box 11: Social Protection Mechanisms

Launched in January 2018, the Pathways for Prosperity Commission on Technology and Inclusive Development is a multistakeholder organization focused on making digital technologies work for the benefit of the world's poorest and most marginalized men and women.[a] Their recent analysis of social protection mechanisms for the digital age highlighted several important insights.[b] The ideal policy will vary from country to country, with different approaches for how to deliver and fund services, which groups to target, and whether obligations should be imposed upon recipients. Targeting payments (for example, to children) is a popular way to ensure that funds only go to those most in need, but screening beneficiaries can be administratively costly (footnote a). Some countries are pursuing universal basic incomes (a flat-rate, unconditional cash transfer to every citizen) to avoid the cost of administrative screening and decision-making. However, the costs of these programs are significant and could be potentially unfeasible for many emerging countries. In terms of funding these programs, there are a variety of options, but all come with potential trade-offs.

[a] Pathways for Prosperity Commission. About. https://pathwayscommission.bsg.ox.ac.uk/inclusion-ready-digital-age.
[b] Pathways for Prosperity Commission. 2019. *The Digital Road map: How Developing Countries Can Get Ahead.* https://pathwayscommission.bsg.ox.ac.uk/sites/default/files/2019-11/the_digital_roadmap.pdf.

Source: Asian Development Bank and AlphaBeta.

Industry-Specific Priorities

While these recommendations apply to both the IT–BPO and electronics manufacturing industries, there are a set of priorities unique to each industry that should be considered when implementing the respective policy actions. Aimed at tackling the underlying weaknesses in each industry's ability to reap the benefits from 4IR technologies, these priorities were based on the findings in the earlier chapters, as well as on in-country consultations with government, industry, and training and education sector stakeholders.

Information Technology and Business Process Outsourcing Industry

- Enhance the effectiveness of current industry–institution engagement efforts. In-country stakeholder consultations reflected that although there was already a high level of engagement between the industry and training and education sector representatives on the industry's skills requirements for 4IR, these efforts had not translated into the curricula shifts needed to close skills gaps. The training institution survey similarly reflected that although 86% of training institutions communicate with employers at least twice a year, less than half of the institutions (46%) review their curricula at least once a year (Chapter 2). With the analysis in Chapter 3 highlighting administrative bottlenecks in the processes to update educational curricula, it is important that government agencies focus on ways to relieve these bottlenecks, particularly for this industry, as outlined in Recommendation 3. Industry-led TVET programs (Recommendation 2) would also be potentially important to addressing these gaps.
- Ensure focus on training critical thinking and complex problem solving skills. The analysis in Chapter 1 reflected that 4IR-related disruptions in the IT–BPO industry are likely to boost the importance of critical thinking skills. As technologies such as AI and chatbots develop and reduce the need for workers to attend to routine tasks (e.g., answering frequently asked questions by customers), more time will be potentially freed up for workers to focus on addressing more complex problems. It is thus critical to ensure that all training programs delivered as part of the recommended actions embed a focus on training for such skills.

- Improve capacity of employers to deliver on-the-job training. With the analysis in Chapter 2 finding that much of the required skills development in response to 4IR in this industry will likely need to take place on the job, it is critical to enhance the capacity of employers to ensure more and higher-quality training in the workplace. This means that greater emphasis will have to be placed on three aspects: increasing employers' awareness of the importance and net benefits of on-the-job training (e.g., through engaging employers in the 4IR transformation road maps as outlined in Recommendation 1), enhancing the financial incentives for employers to do this (e.g., through implementing the incentive schemes outlined in Recommendation 6), and increasing their own capacity and adaptiveness to 4IR technologies (e.g., by developing flexible training programs for employers themselves as outlined in Recommendation 5).

Electronics Manufacturing Industry

- Support 4IR knowledge transfer from large multinational companies to micro, small, and medium-sized enterprises (MSMEs). The Philippines' electronics manufacturing industry is an ecosystem consisting of the local entities of large multinational companies (e.g., TI Philippines, Toshiba Information Equipment (Philippines) Inc., Fairchild Semiconductor Philippines Inc.), and small and medium-sized parts and components manufacturing companies that typically act as suppliers to the larger firms.[65] Being better resourced and with stronger international networks, the large companies are generally in a more advanced stage of 4IR adoption and training than MSMEs. A key concern reflected by government and industry representatives during in-country stakeholder consultations was that MSMEs would lag in terms of their 4IR preparedness. There is thus a compelling push for there to be knowledge transfer of 4IR adoption and skills development strategies from the large companies to MSMEs. These large companies can also be instrumental in supporting many of the policy recommendations outlined—from formulating the 4IR transformation road maps for the electronics manufacturing industry (Recommendation 1) to co-creating the skills certification programs for the industry (Recommendation 5) that MSMEs can then incorporate into their own training programs. Where the required competencies overlap, there could even be scope for consolidated training programs led by these large companies for the benefit of MSMEs.
- Address the potentially disproportionate impact of technological disruption on women. The analysis in Chapter 1 reflected that the automation impact in this industry was likely to be felt mainly on manual jobs, which tend to be held mostly by women. It is thus important that the training programs described in the recommendations incorporate gender-sensitive approaches (e.g., Recommendation 2 on industry-led TVET programs; Recommendation 5 on flexible skills certification programs; Recommendation 7 on the formulation of new training approaches for vulnerable workers). These could include consideration of teaching methods that have been demonstrated to be more effective for female learners (e.g., having female STEM role models as trainers).[66]
- Develop a standardized set of 4IR skills requirements and training quality standards. In-country consultations with industry stakeholders showed that while there were current partnerships forged between electronics firms and educational and training institutions to improve the relevance of the latter's curricula, these tended to take place on an ad hoc basis and were sporadic throughout the industry as well as across education and training institutions. This has led to uncertainty about the standard required of workers in the industry, as well as unevenness

---

[65] Government of the Philippines, Gov.ph. About electronics. http://industry.gov.ph/industry/electronics/.
[66] Microsoft. 2018. *Closing the STEM Gap: Why STEM Classes and Careers Still Lack Girls and What We Can Do About It.* https://query.prod.cms.rt.microsoft.com/cms/api/am/binary/RE1UMWz.

in training and education curricula across institutions (which depend largely on the industry partners they worked with). As result, there has been a strong call by industry for the government to standardize skills requirements for the industry[67] (which can be achieved through the 4IR transformation road maps outlined in Recommendation 1), and to enforce them across all training programs (as in the modular skills certification programs outlined in Recommendation 5).

---

[67]   Semiconductors and Electronics Industries in the Philippines, Inc. Undated. The Philippine Semiconductor and Electronics Industry Road map. http://industry.gov.ph/wp-content/uploads/2015/05/4th-TID-Mr.-Lachicas-Presentation-on-Electronics.pdf.

List of Participants Engaged During National Consultations

Together with the Asian Development Bank (ADB), the AlphaBeta team consulted a range of government and industry stakeholders in a series of initial consultations in July 2019, and subsequently through a workshop in October 2019. Refer to Tables A1 and A2 for the lists of consulted stakeholders in both engagements.

Table A1: List of Stakeholders Engaged in Initial Consultations in July 2019

Entity	Stakeholders Engaged
Government Agencies	
Technical Education and Skills Development Authority (TESDA)	• AED Rosalina S. Constantino, Assistant Executive Director, Planning Office (PO) • Celestino C. Millar, Chief Technical Education and Skills Development (TESD) Specialist, Policies and Planning Division (PPD), Planning Office (PO) • Regino C. Cleofe, Senior TESDS, Policy Research and Evaluation Division (PRED), PO • Christian Andrei C. Aguilar, TESDS II, Project Development Division (PDD), PO • Stephen Cezar, Senior TESD Specialist
Department of Labor and Employment (DOLE)	• Dominique Rubia-Tutay, Director IV, Bureau of Local Employment • Myka Rose Trono, Statistician II • Precious Nicole Bugayong, Labor and Employment Officer II
Department of Trade and Industry (DTI)	• Rhea Matute, Executive Director, Design Center of the Philippines • Paula Celis, Project Officer, Design Center of the Philippines • Daniele Baduria, Senior Industrial Design Specialist, Design Center of the Philippines • Abigail R. Zurita, Assistant Director of the Bureau of Trade and Industrial Policy Research (BHPR) • Myleen V. Aldana, Division Chief of BHPR
Industry Associations and Major Employers	
IT & Business Process Association of the Philippines (IBPAP)	• Carole Gaffud, Senior Research Manager • Zoe Diaz De Rivera, Master Trainer
Makati Business Club	• Nova Nguyen
Philippine Constructors Association	• Earl Segales

continued on next page

Table A1 *continued*

Entity	Stakeholders Engaged
Employer Groups and Major Employers	
Employers Confederation of the Philippines	• Antonio Ll. Sayo, Governor & Chairperson of the Technical Working Group on Skills Development • Renato B. Almeda, Governor • Jose Roland Moya, Deputy Director-General
EMS Components Assembly Inc.	• Dennis Caronan
Sitel Philippines Inc.	• Haidee Enriquez, General Manager
Education, Training and Career Development Institutions	
De La Salle University Manila	• Kris Francisco, Associate Professor
Don Bosco Training Institute	• Neil Apollo S. Dichoso, National Compliance Officer
DualTech Training Center	• Marvin Adolfo, Corporate Secretary
SFI Career Center	• Gina A. Jusay, Managing Director • Richard Monteverde, Senior Program Officer • Danielle Marie Cruz, Career Coach and Program Associate
Career Development Association of the Philippines	• Jose Alejo, Vice President
Others (Research Institutions, Multilateral Organizations)	
ILO Philippines Country Office	• Khalid Hassan, Director • Linartes "Lites" M. Viloria, National Project Coordinator for the Women in STEM Workforce Readiness Program
Philippine Institute for Development Studies	• Jose Ramon G. Albert, Fellow II

Source: Asian Development Bank and AlphaBeta.

Table A2: List of Stakeholders Engaged in In-Country Workshop in October 2019

Entity	Stakeholders Engaged
Government Agencies	
Commission on Higher Education (CHED)	• Cherrie Melanie Ancheta-Diego, Director
Department of Education (DepEd)	• Alma Torio, Assistant Secretary for Curriculum and Instruction • Margarita Galias, Technical Assistant • Jose Adrian C. Fernandez, Executive Assistant IV

continued on next page

Table A2 *continued*

Entity	Stakeholders Engaged
Department of Finance (DOF)	• Nheza May Loquinte, Project Evaluation Officer II • Christine Seana Verano, Budget and Management Specialist II • Jason Angelo Trabuco, Budget and Management Specialist II
Department of Information and Communication Technology (DICT)	• Malou Aquilizan, Director • Aida Yuvienco, Director
Department of Labor and Employment (DOLE)	• Myka Rose Trono, Statistician II2 • Andrea Joi Adorio, JobStart Project Coordinator
Department of Trade and Industry (DTI)	• Abigail Zurita, Assistant Director • Elli May N. Malabayabas, Trade-Industry Development • Myleen V. Aldana, Division Chief
National Economic and Development Authority (NEDA)	• Susan Carandang, Supervising Economic Development Specialist
Technical Education Skills and Development Authority (TESDA)	• Katherine Amor A. Zarsadias, Chief TESDA Specialist, Policy Research Evaluation Division Planning Office • Ma. Lina A. Andrade, Supervising TESD Specialist • Rosalina S. Constantino, Assistant Executive Director of Planning Office
Industry	
IT & Business Process Association (IBPAP)	• Carole Gaffud, Senior Research Manager • Frankie Antolin, Executive Director
Philippine Chamber of Commerce and Industry	• Raymund Margallo, Project Officer • Patricia de la Chica, Project Senior Staff
Philippine Exporters Confederation Inc. (Philexport)	• Liza Leong, Vice President • Andrea Dabuet, Policy Officer
Semiconductor electronics Industry Foundation Inc.	• Danny Javid
Training and Career Development Institutions	
Don Bosco – One TVET Philippines	• Jose Dindo S. Vitug, Executive Director
Dualtech Training Institute	• Marvin Adolfo, Corporate Secretary
SFI Career Center	• Gina A. Jusay, Managing Director • Richard P. Moteverder, Programs Officer • Daniel Marie Cruz, Programs Specialist

Source: Asian Development Bank and AlphaBeta.

Bibliography

Accenture. 2012. *Utilizing Analytics to Maximize Business Outcomes*. https://www.accenture.com/us-en/~/media/accenture/conversion-assets/dotcom/documents/global/pdf/technology_7/accenture-utilizing-analytics-maximize-business-outcomes.pdf.

ACT/EMP & International Labour Organization. 2017. *ASEAN in Transformation—How Technology is Changing Jobs and Enterprises—The Philippines Country Brief*. https://www.ilo.org/actemp/publications/WCMS_579667/lang--en/index.htm.

Asian Development Bank (ADB). 2018. *Asian Development Outlook 2018: How Technology Affects Jobs*. https://www.adb.org/publications/asian-development-outlook-2018-how-technology-affects-jobs.

ADB. 2018. *Social Protection Brief: Reducing Youth Not in Employment, Education or Training through JobStart Philippines*. https://www.adb.org/sites/default/files/publication/396081/adb-brief-084-youth-not-employment.pdf.

Alferez, R. C. 2012. Implementation of Strengthened Technical Vocational Education Program—Competency Based Curriculum, Northern Mindanao, Philippines. *JPAIR Multidisciplinary Research Journal*. 7 (1). https://ejournals.ph/article.php?id=7497.

AlphaBeta. 2017. *The Automation Advantage*. https://www.alphabeta.com/our-research/the-automation-advantage/.

Aquino, L. A. 2019. Youth unemployment still dominant problem in labor market – DOLE. Manila Bulletin. 1 January. http://openstat.psa.gov.ph/Metadata/3E3D6080.

Arbulu, I. et al. 2018. *Industry 4.0: Reinvigorating ASEAN Manufacturing for the Future*. McKinsey & Company. https://www.mckinsey.com/business-functions/operations/our-insights/industry-4-0-reinvigorating-asean-manufacturing-for-the-future.

Asia Philanthropy Circle. 2017. *Catalysing Productive Livelihood: A Guide to Education Interventions with an Accelerated Path to Scale and Impact*. http://www.edumap-indonesia.asiaphilanthropycircle.org/wp-content/uploads/2017/11/APC-Giving-Guide-Book-Final-Report-17112017.pdf.

Barber, M. 2007. *Instruction to Deliver: Fighting to Transform Britain's Public Services*: Methuen Publishing Ltd.

The Behavioural Insights Team, Cabinet Office, and Nesta. 2015. *Easy, Attractive, Timely, Social: Four Simple Ways to Apply Behavioural Insights.* https://www.behaviouralinsights.co.uk/wp-content/uploads/2015/07/BIT-Publication-EAST_FA_WEB.pdf.

Bonoli, G. 2019. Ensuring economic security in the gig economy. *The Business Times.* 13 March. https://www.businesstimes.com.sg/opinion/ensuring-economic-security-in-the-gig-economy.

Budhrani, K. S. et al. 2017. *Developing a Skilled Workforce Through Technical and Vocational Education and Training in the Philippines.* https://link.springer.com/referenceworkentry/10.1007%2F978-3-319-38909-7_28-1.

Center for Universal Education at Brookings. 2019. *Learning to Leapfrog: Innovative Pedagogies to Transform Education.* https://www.brookings.edu/wp-content/uploads/2019/09/Learning-to-Leapfrog-InnovativePedagogiestoTransformEducation-Web.pdf.

Cognizant Center for the Future of Work. 2017. *21 Jobs of the Future: A Guide to Getting—and Staying—Employed for the Next 10 Years.* https://www.cognizant.com/whitepapers/21-jobs-of-the-future-a-guide-to-getting-and-staying-employed-over-the-next-10-years-codex3049.pdf.

Cornell University, INSEAD, and World Intellectual Property Organization. 2019. *Global Innovation Index 2019: Creating Healthy Lives—The Future of Medical Innovation.* https://www.wipo.int/edocs/pubdocs/en/wipo_pub_gii_2019.pdf.

Daly, E. and S. Singham. 2012. Delivery 2.0: The New Challenge for Governments. *McKinsey & Company.* 1 September. https://www.mckinsey.com/industries/public-sector/our-insights/delivery-20-the-new-challenge-for-governments.

Gartner Research. 2018. *Market Guide for Conversational Platforms.* https://www.gartner.com/en/documents/3879492.

Generation. https://www.generation.org/.

Government of Malaysia, Department of Skills Development. Malaysian Skill Certificate (SKM). https://www.dsd.gov.my/index.php/en/service/malaysian-skills-certificate.

Government of Malaysia, Ministry of Higher Education and Malaysian Qualifications Agency. 2011. Malaysian Qualifications Framework. https://www.mqa.gov.my/pv4/mqf.cfm.

Government of the Philippines, Department of Budget and Management. 2019. TESDA Budget Nearly Doubles in 2019. 17 October. https://www.dbm.gov.ph/index.php/secretary-s-corner/press-releases/list-of-press-releases/1247-tesda-budget-nearly-doubles-in-2019.

Governmentof the Philippines, Department of Information and Communications Technology. 2017. *National Broadband Plan.* https://dict.gov.ph/wp-content/uploads/2017/09/2017.08.09-National-Broadband-Plan.pdf.

Government of the Philippines, Department of Labor and Employment (DOLE). 2019. *JobsFit 2022—Labor Market Information Report*. http://www.ble.dole.gov.ph/index.php/web-pages/120-jobsfit.

———. JobStart Philippines Program. https://www.dole.gov.ph/jobstart-philippines-program/.

———. 2017. *Harmonized National Research and Development Agenda 2017–2022*. http://dost.gov.ph/phocadownload/Downloads/Journals/Approved%20Harmonized%20National%20RD%20Agenda%20%202017-2022.pdf.

Government of the Philippines, Department of Science and Technology (DOST). 2018. *Guidelines for the Accelerated R&D Program for Capacity Building of Research and Development Institutions and Industrial Competitiveness of the Science for Change (S4C) Program*. http://www.dost.gov.ph/phocadownload/Downloads/Resources/SCIENCE_FOR_CHANGE_PROGRAM_S4CP/S4CP_Guidelines_07Feb2018.pdf.

———. 2019. Science for Change Program (SC4P). http://www.dost.gov.ph/knowledge-resources/downloads/category/119-science-for-change-program-s4cp.html.

———. 2019. SETUP (Small Enterprise Technology Upgrading Program). http://region7.dost.gov.ph/programs/technology-transfer-and-commercialization/setup/.

Government of the Philippines, Department of Trade and Industry (DTI). 2017. *Philippine Inclusive Innovation Industrial Strategy*. http://boi.gov.ph/sdm_downloads/dti-policy-brief-2017-05-philippine-inclusive-innovation-industrial-strategy/.

Government of the Philippines, DTI, and Philippine Board of Investments (BOI). 2012. *Comprehensive National Industrial Strategy (CNIS)*. http://industry.gov.ph/comprehensive-national-industrial-strategy/.

Government of the Philippines, DTI, and BOI. Undated. *Industry 4.0: Are We There Yet? I³S Inclusive Innovation Industrial Strategy*. https://pidswebs.pids.gov.ph/CDN/EVENTS/industrialstrategy_aldaba.pdf.

Government of the Philippines, Gov.ph. About electronics. http://industry.gov.ph/industry/electronics/.

Government of the Philippines, National Economic and Development Authority. 2017. *Philippine Development Plan 2017–2022*. http://pdp.neda.gov.ph/#:~:text=The%20Philippine%20Development%20Plan%202017,NA%20BUHAY%20PARA%20SA%20LAHAT.

Government of the Philippines, Philippine Board of Investments (BOI) and IT and Business Process Association of the Philippines. 2018. *Accelerate PH: Future Ready Road map 2022*. https://boi.gov.ph/wp-content/uploads/2018/03/Executive-Summary-Accelerate-PH-Future-Ready-Roadmap-2022_with-corrections.pdf.

Government of the Philippines, Philippines Statistics Authority. 2018. *Filipino Families Are Most Deprived in Education*. https://psa.gov.ph/sites/default/files/mpi%20press%20release.pdf.

Government of the Philippines, Technical Education and Skills Development Authority (TESDA). 2018. *National Technical Education and Skills Development Plan (2018–2022)*. http://www.tesda.gov.ph/About/TESDA/47.

Government of the Philippines, TESDA. 2019. Scholarship and Student Assistance Programs. http://www.tesda.gov.ph/About/TESDA/1279.

Government of Singapore, Ministry of Education. 2016. *Formation of the Council for Skills, Innovation and Productivity*. 20 May. https://www.moe.gov.sg/news/press-releases/formation-of-the-council-for-skills--innovation-and-productivity.

Government of Singapore, Ministry of Trade and Industry (MTI). 2017. *Media Factsheet—Industry Transformation Maps: Integrated Road maps to Drive Industry Transformation*. https://www.mti.gov.sg/-/media/MTI/ITM/General/Fact-sheet-on-Industry-Transformation-Maps---revised-as-of-31-Mar-17.pdf.

Government of Singapore, MTI. 2020. The Future Economy Council. https://www.mti.gov.sg/FutureEconomy/TheFutureEconomyCouncil.

Hardy, W. et al. 2019. Technology, Skills And Globalization: Explaining International Differences in Routine and Non-Routine Work Using Survey Data. IBS working paper 04/2019. https://ibs.org.pl/en/publications/technology-skills-and-globalization-explaining-international-differences-in-routine-and-nonroutine-work-using-survey-data/.

Hasnan, L. 2019. Philippines's Fast-Growing Gig Economy. *The ASEAN Post*. 9 October. https://theaseanpost.com/article/philippines-fast-growing-gig-economy.

IL&FS. 2019. *Our Work: Skill Development*. https://www.ilfsindia.com/our-work/skill-development/.

International Labour Organization (ILO). 2016. *ASEAN in Transformation: How Technology is Changing Jobs and Enterprises*. https://unctad.org/meetings/en/Presentation/cstd2016_p24_Jae-HeeChang_ILO_en.pdf.

ILO. 2020. Philippines—Dual Training System Act of 1994. https://www.ilo.org/dyn/natlex/natlex4.detail?p_lang=&p_isn=38359&p_classification=09.

International Federation of Robotics. 2018. Automation Boom In Electrical/Electronics Industry Drives 30% Increase in Sales of Industrial Robots. 7 November. https://ifr.org/post/automation-boom-in-electrical-electronics-industry-drives-30-increase-in-sa.

Ketelhut, D. et al. 2009. A Multi-User Virtual Environment for Building and Assessing Higher Order Inquiry Skills in Science. *British Journal of Educational Technology*. https://onlinelibrary.wiley.com/doi/abs/10.1111/j.1467-8535.2009.01036.x.

KellyOCG. 2018. From Workforce to Workfit. https://www.kellyocg.com/insights/featured-content/whitepapers/from-workforce-to-workfit/.

Kim, J. et al. 2019. Philippine Readiness for the 4th Industrial Revolution: A Case Study. *Asia-Pacific Social Science Review*. http://apssr.com/wp-content/uploads/2019/03/RA-9-R.pdf.

Lorenzo, M. P. M. 2017. JobStart Philippines: A Promising Project with Some Obstacles. 28 August. https://policyblog.uni-graz.at/2017/08/jobstart-philippines-a-promising-project-with-some-obstacles/.

Macha, W. et al. 2018. Education System Profiles—Education in the Philippines. https://wenr.wes. org/2018/03/education-in-the-philippines.

Maclean, R. et al., eds. 2012. *Skills Development for Inclusive and Sustainable Growth in Developing Asia-Pacific*. https://www.adb.org/publications/skills-development-inclusive-and-sustainable-growth-developing-asia-pacific.

Microsoft. 2018. *The Future Computed: Artificial Intelligence and its Role in Society*. https://blogs.microsoft. com/wp-content/uploads/2018/02/The-Future-Computed_2.8.18.pdf.

Microsoft. 2018. *Closing the STEM Gap: Why STEM Classes and Careers Still Lack Girls and What We Can Do About It*. https://query.prod.cms.rt.microsoft.com/cms/api/am/binary/RE1UMWz.

Microsoft Asia Stories. 2019. *Preparing for AI: The Implications of Artificial Intelligence for Jobs and Skills in Asian Economies*. https://news.microsoft.com/apac/2019/08/26/preparing-for-ai-the-implications-of-artificial-intelligence-for-jobs-and-skills-in-asian-economies/.

Microsoft News Center Philippines. 2018. *Digital Transformation to Contribute US$8 billion to the Philippines GDP by 2021*. https://news.microsoft.com/en-ph/2018/02/14/digital-transformation-contribute-us8-billion-philippines-gdp-2021/.

Olson, P. 2018. Google, Microsoft And Startups Are Going To War On Chatbot Technology. *Forbes*. 27 July. https://www.forbes.com/sites/parmyolson/2018/07/27/google-microsoft-and-startups-are-going-to-war-on-chatbot-technology/#d9229c561b67.

O'Malley, M. 2018. *PayPal Releases Global Freelancer Insights*. https://www.paypal.com/stories/us/paypal-releases-global-freelancer-insights.

onebillion. Undated. *onecourse: one app that delivers reading, writing and numeracy*. https://onebillion.org/onecourse/app/.

Orbeta Jr., A. C. et al. 2016. *Are Higher Education Institutions Responsive to Changes in the Labor Market?* https://dirp4.pids.gov.ph/websitecms/CDN/PUBLICATIONS/pidsdps1608.pdf.

Organisation for Economic Co-operation and Development (OECD). 2010. *Learning for Jobs—The OECD International Survey of VET Systems: First Results and Technical Report*. https://www.oecd.org/education/skills-beyond-school/47334855.pdf.

———. 2012. *Skills Development Pathways in Asia*. https://www.oecd.org/cfe/leed/Skills%20 Development%20Pathways%20in%20Asia_FINAL%20VERSION.pdf.

———. 2017. *Employment and Skills Strategies in the Philippines*. https://www.oecd.org/publications/employment-and-skills-strategies-in-the-philippines-9789264273436-en.htm.

———. 2018. *The Future of Social Protection: What Works for Non-Standard Workers?* https://doi. org/10.1787/9789264306943-en.

Oxford Economics. 2018. *Technology and the future of ASEAN jobs: The Impact of AI on Workers in ASEAN's Six Largest Economies*. https://www.oxfordeconomics.com/recent-releases/dd577680-7297-4677-aa8f-450da197e132.

Oxford Internet Institute. 2019. The Online Labour Index. https://ilabour.oii.ox.ac.uk/online-labour-index/.

Pathways for Prosperity Commission. About. https://pathwayscommission.bsg.ox.ac.uk/inclusion-ready-digital-age.

Pathways for Prosperity Commission. 2019. *Positive Disruption: Health and Education in the Digital Age*. https://pathwayscommission.bsg.ox.ac.uk/positive-disruption.

Pathways for Prosperity Commission. 2019. *The Digital Road map: How Developing Countries Can Get Ahead*. https://pathwayscommission.bsg.ox.ac.uk/sites/default/files/2019-11/the_digital_roadmap.pdf.

Payoneer. 2019. *The Global Gig-Economy Index: Q2 2019*. https://explore.payoneer.com/q2_global_freelancing_index/.

Philippine Institute for Development Studies. 2018. *Preparing the Philippines for the Fourth Industrial Revolution: A Scoping Study*. https://pidswebs.pids.gov.ph/CDN/PUBLICATIONS/pidsdps1811.pdf.

Prospera & AlphaBeta Advisors. 2019. *Capturing Indonesia's Automation Potential*. https://www.alphabeta.com/wp-content/uploads/2019/08/capturing-indonesias-automation-potential.pdf.

Schwab, K. 2017. *The Fourth Industrial Revolution*. https://books.google.com.sg/books?hl=en&lr=&id=ST_FDAAAQBAJ&oi=fnd&pg=PR7&dq=klaus+schwab+fourth+industrial+revolution&ots=DTnvbTqvTQ&sig=aOLqcUCFsLKbNpjWa5kr2Sjzhu4#v=onepage&q=klaus%20schwab%20fourth%20industrial%20revolution&f=false.

Semiconductors and Electronics Industries in the Philippines, Inc. Undated. The Philippine Semiconductor and Electronics Road map. http://industry.gov.ph/wp-content/uploads/2015/05/4th-TID-Mr.-Lachicas-Presentation-on-Electronics.pdf.

Sharp, N. 2019. *Is Additive Manufacturing the Right Choice for Your Electronic Assembly?* 7 November. https://blog.jjsmanufacturing.com/additive-manufacturing-electronic-assembly.

SkillsFuture SG. 2019. Funding Support for Employers. https://www.ssg.gov.sg/programmes-and-initiatives/funding/funding-for-employer-based-training.html.

SkillsFuture. 2019. Skills Framework. https://www.skillsfuture.sg/skills-framework.

Institute of Marketing Malaysia. 2019. SMECorp's skills upgrading programme. https://imm.org.my/smecorps-skills-upgrading-programme.

Spector, J. M. and S. Ma. 2019. Inquiry and critical thinking skills for the next generation: from artificial intelligence back to human intelligence. *Smart Learning Environments*. 9 (8). https://slejournal.springeropen.com/articles/10.1186/s40561-019-0088-z.

Tang, S. K. 2019. Singapore businesses not investing enough in employee training: SBF survey. *Channel News Asia.* 17 January. https://www.channelnewsasia.com/news/business/singapore-companies-not-investing-employee-training-sbf-survey-11134230.

United Nations Industrial Development Organization. 2018. *You Say You Want a Revolution: Strategic Approaches to Industry 4.0 in Middle-Income Countries.* https://www.unido.org/api/opentext/documents/download/10031392/unido-file-10031392.

Veal, K. 2017. *Partnering with Industry: Employer and Institute Linkages.* https://development.asia/explainer/partnering-industry-employer-and-institute-linkages.

Woetzl, J. et al. 2014. *Southeast Asia at the Crossroads: Three Paths to Prosperity.* McKinsey Global Institute. November. https://www.mckinsey.com/~/media/McKinsey/Featured%20Insights/Asia%20Pacific/Three%20paths%20to%20sustained%20economic%20growth%20in%20Southeast%20Asia/MGI%20SE%20Asia_Executive%20summary_November%202014.ashx.

World Bank. The STEP Skills Measurement Program. https://microdata.worldbank.org/index.php/catalog/step/about.

World Economic Forum and A. T. Kearney. 2018. *Readiness for the Future of Production Report 2018.* http://www3.weforum.org/docs/FOP_Readiness_Report_2018.pdf.

World Economic Forum. 2018. *The Future of Jobs Report 2018.* http://www3.weforum.org/docs/WEF_Future_of_Jobs_2018.pdf.

World Economic Forum and Boston Consulting Group. 2019. *Towards a Reskilling Revolution: Industry-Led Action for the Future of Work.* http://www3.weforum.org/docs/WEF_Towards_a_Reskilling_Revolution.pdf.